NATURAL HISTORY
UNIVERSAL LIBRARY

西方博物学大系

主编：江晓原

NATURAL HISTORY OF THE INSECTS OF CHINA
NATURAL HISTORY OF THE INSECTS OF INDIA

中国昆虫志
印度昆虫志

[英] 爱德华·多诺万 著

华东师范大学出版社

图书在版编目（CIP）数据

中国昆虫志·印度昆虫志 = Natural history of the insects of China · Natural history of the insects of India：英文 /（英）爱德华·多诺万著. — 上海：华东师范大学出版社, 2018
（寰宇文献）
ISBN 978-7-5675-7721-3

Ⅰ.①中… Ⅱ.①爱… Ⅲ.①昆虫志-中国-英文
Ⅳ.①Q968.22 ②Q968.235.1

中国版本图书馆CIP数据核字(2018)第094129号

中国昆虫志·印度昆虫志
Natural history of the insects of China · Natural history of the insects of India
（英）爱德华·多诺万著

特约策划	黄曙辉　徐　辰
责任编辑	庞　坚
特约编辑	许　倩
装帧设计	刘怡霖

出版发行	华东师范大学出版社
社　　址	上海市中山北路3663号　邮编 200062
网　　址	www.ecnupress.com.cn
电　　话	021-60821666　行政传真　021-62572105
客服电话	021-62865537
门市（邮购）电话	021-62869887
地　　址	上海市中山北路3663号华东师范大学校内先锋路口
网　　店	http://hdsdcbs.tmall.com/

印 刷 者	虎彩印艺股份有限公司
开　　本	16开
印　　张	22.5
版　　次	2018年6月第1版
印　　次	2018年6月第1次
书　　号	ISBN 978-7-5675-7721-3
定　　价	560.00元（精装全一册）

出版人　王　焰

（如发现本版图书有印订质量问题，请寄回本社客服中心调换或电话021-62865537联系）

总　目

《西方博物学大系总序》（江晓原）	1
出版说明	1
Natural History of the Insect of China	1
Natural History of the Insect of India	177

《西方博物学大系》总序

江晓原

《西方博物学大系》收录博物学著作超过一百种,时间跨度为 15 世纪至 1919 年,作者分布于 16 个国家,写作语种有英语、法语、拉丁语、德语、弗莱芒语等,涉及对象包括植物、昆虫、软体动物、两栖动物、爬行动物、哺乳动物、鸟类和人类等,西方博物学史上的经典著作大备于此编。

中西方"博物"传统及观念之异同

今天中文里的"博物学"一词,学者们认为对应的英语词汇是 Natural History,考其本义,在中国传统文化中并无现成对应词汇。在中国传统文化中原有"博物"一词,与"自然史"当然并不精确相同,甚至还有着相当大的区别,但是在"搜集自然界的物品"这种最原始的意义上,两者确实也大有相通之处,故以"博物学"对译 Natural History 一词,大体仍属可取,而且已被广泛接受。

已故科学史前辈刘祖慰教授尝言:古代中国人处理知识,如开中药铺,有数十上百小抽屉,将百药分门别类放入其中,即心安矣。刘教授言此,其辞若有憾焉——认为中国人不致力于寻求世界"所以然之理",故不如西方之分析传统优越。然而古代中国人这种处理知识的风格,正与西方的博物学相通。

与此相对,西方的分析传统致力于探求各种现象和物体之间的相互关系,试图以此解释宇宙运行的原因。自古希腊开始,西方哲人即孜孜不倦建构各种几何模型,欲用以说明宇宙如何运行,其中最典型的代表,即为托勒密(Ptolemy)的宇宙体系。

比较两者,差别即在于:古代中国人主要关心外部世界"如何"运行,而以希腊为源头的西方知识传统(西方并非没有别的知识传统,只是未能光大而已)更关心世界"为何"如此运行。在线性发展无限进步的科学主义观念体系中，我们习惯于认为"为何"是在解决了"如何"之后的更高境界，故西方的分析传统比中国的传统更高明。

然而考之古代实际情形，如此简单的优劣结论未必能够成立。例如以天文学言之，古代东西方世界天文学的终极问题是共同的：给定任意地点和时刻，计算出太阳、月亮和五大行星（七政）的位置。古代中国人虽不致力于建立几何模型去解释七政"为何"如此运行，但他们用抽象的周期叠加（古代巴比伦也使用类似方法），同样能在足够高的精度上计算并预报任意给定地点和时刻的七政位置。而通过持续观察天象变化以统计、收集各种天象周期，同样可视之为富有博物学色彩的活动。

还有一点需要注意：虽然我们已经接受了用"博物学"来对译 Natural History，但中国的博物传统，确实和西方的博物学有一个重大差别——即中国的博物传统是可以容纳怪力乱神的，而西方的博物学基本上没有怪力乱神的位置。

古代中国人的博物传统不限于"多识于鸟兽草木之名"。体现此种传统的典型著作，首推晋代张华《博物志》一书。书名"博物"，其义尽显。此书从内容到分类，无不充分体现它作为中国博物传统的代表资格。

《博物志》中内容，大致可分为五类：一、山川地理知识；二、奇禽异兽描述；三、古代神话材料；四、历史人物传说；五、神仙方伎故事。这五大类，完全符合中国文化中的博物传统，深合中国古代博物传统之旨。第一类，其中涉及宇宙学说，甚至还有"地动"思想，故为科学史家所重视。第二类，其中甚至出现了中国古代长期流传的"守宫砂"传说的早期文献：相传守宫砂点在处女胳膊上，永不褪色，只有性交之后才会自动消失。第三类，古代神话传说，其中甚至包括可猜想为现代"连体人"的记载。第四类，各种著名历史人物，比如三位著名刺客的传说，此三名刺客及所刺对象，历史上皆实有其人。第五类，包括各种古代方术传说，比如中国古代房中养生学说，房中术史上的传说人物之一"青牛道士封君达"等等。前两类与西方的博物学较为接近，但每一类都会带怪力乱神色彩。

"所有的科学不是物理学就是集邮"

在许多人心目中，画画花草图案，做做昆虫标本，拍拍植物照片，这类博物学活动，和精密的数理科学，比如天文学、物理学等等，那是无法同日而语的。博物学显得那么的初级、简单，甚至幼稚。这种观念，实际上是将"数理程度"作为唯一的标尺，用来衡量一切知识。但凡能够使用数学工具来描述的，或能够进行物理实验的，那就是"硬"科学。使用的数学工具越高深越复杂，似乎就越"硬"；物理实验设备越庞大，花费的金钱越多，似乎就越"高端"、越"先进"……

这样的观念，当然带着浓厚的"物理学沙文主义"色彩，在很多情况下是不正确的。而实际上，即使我们暂且同意上述"物理学沙文主义"的观念，博物学的"科学地位"也仍然可以保住。作为一个学天体物理专业出身，因而经常徜徉在"物理学沙文主义"幻影之下的人，我很乐意指出这样一个事实：现代天文学家们的研究工作中，仍然有绘制星图，编制星表，以及为此进行的巡天观测等等活动，这些活动和博物学家"寻花问柳"，绘制植物或昆虫图谱，本质上是完全一致的。

这里我们不妨重温物理学家卢瑟福（Ernest Rutherford）的金句："所有的科学不是物理学就是集邮（All science is either physics or stamp collecting）。"卢瑟福的这个金句堪称"物理学沙文主义"的极致，连天文学也没被他放在眼里。不过，按照中国传统的"博物"理念，集邮毫无疑问应该是博物学的一部分——尽管古代并没有邮票。卢瑟福的金句也可以从另一个角度来解读：既然在卢瑟福眼里天文学和博物学都只是"集邮"，那岂不就可以将博物学和天文学相提并论了？

如果我们摆脱了科学主义的语境，则西方模式的优越性将进一步被消解。例如，按照霍金（Stephen Hawking）在《大设计》（The Grand Design）中的意见，他所认同的是一种"依赖模型的实在论（model-dependent realism）"，即"不存在与图像或理论无关的实在性概念（There is no picture- or theory-independent concept of reality）"。在这样的认识中，我们以前所坚信的外部世界的客观性，已经不复存在。既然几何模型只不过是对外部世界图像的人为建构，则古代中国人干脆放弃这种建构直奔应用（毕竟在实际应用中我们只需要知道七政"如何"运行),又有何不可?

传说中的"神农尝百草"故事,也可以在类似意义下得到新的解读:"尝百草"当然是富有博物学色彩的活动,神农通过这一活动,得知哪些草能够治病,哪些不能,然而在这个传说中,神农显然没有致力于解释"为何"某些草能够治病而另一些则不能,更不会去建立"模型"以说明之。

"帝国科学"的原罪

今日学者有倡言"博物学复兴"者,用意可有多种,诸如缓解压力、亲近自然、保护环境、绿色生活、可持续发展、科学主义解毒剂等等,皆属美善。编印《西方博物学大系》也是意欲为"博物学复兴"添一助力。

然而,对于这些博物学著作,有一点似乎从未见学者指出过,而鄙意以为,当我们披阅把玩欣赏这些著作时,意识到这一点是必须的。

这百余种著作的时间跨度为15世纪至1919年,注意这个时间跨度,正是西方列强"帝国科学"大行其道的时代。遥想当年,帝国的科学家们乘上帝国的军舰——达尔文在皇家海军"小猎犬号"上就是这样的场景之一,前往那些已经成为帝国的殖民地或还未成为殖民地的"未开化"的遥远地方,通常都是踌躇满志、充满优越感的。

作为一个典型的例子,英国学者法拉在(Patricia Fara)《性、植物学与帝国:林奈与班克斯》(*Sex, Botany and Empire, The Story of Carl Linnaeus and Joseph Banks*)一书中讲述了英国植物学家班克斯(Joseph Banks)的故事。1768年8月15日,班克斯告别未婚妻,登上了澳大利亚军舰"奋进号"。此次"奋进号"的远航是受英国海军部和皇家学会资助,目的是前往南太平洋的塔希提岛(Tahiti,法属海外自治领,另一个常见的译名是"大溪地")观测一次比较罕见的金星凌日。舰长库克(James Cook)是西方殖民史上最著名的舰长之一,多次远航探险,开拓海外殖民地。他还被认为是澳大利亚和夏威夷群岛的"发现"者,如今以他命名的群岛、海峡、山峰等不胜枚举。

当"奋进号"停靠塔希提岛时,班克斯一下就被当地美丽的土著女性迷昏了，他在她们的温柔乡里纵情狂欢，连库克舰长都看不下去了，"道德愤怒情绪偷偷溜进了他的日志当中，他发现自己根本不可能不去批评所见到的滥交行为"，而班克斯纵欲到了"连嫖妓都毫无激情"的地步——这是别人讽刺班克斯的说法，因为对于那时常年航行于茫茫大海上的男性来说，上岸嫖妓通常是一项能够唤起"激情"的活动。

而在"帝国科学"的宏大叙事中，科学家的私德是无关紧要的，人们关注的是科学家做出的科学发现。所以，尽管一面是班克斯在塔希提岛纵欲滥交，一面是他留在故乡的未婚妻正泪眼婆娑地"为远去的心上人绣织背心"，这样典型的"渣男"行径要是放在今天，非被互联网上的口水淹死不可，但是"班克斯很快从他们的分离之苦中走了出来，在外近三年，他活得倒十分滋润"。

法拉不无讽刺地指出了"帝国科学"的实质："班克斯接管了当地的女性和植物，而库克则保护了大英帝国在太平洋上的殖民地。"甚至对班克斯的植物学本身也调侃了一番："即使是植物学方面的科学术语也充满了性指涉。……这个体系主要依靠花朵之中雌雄生殖器官的数量来进行分类。"据说"要保护年轻妇女不受植物学教育的浸染，他们严令禁止各种各样的植物采集探险活动。"这简直就是将植物学看成一种"涉黄"的淫秽色情活动了。

在意识形态强烈影响着我们学术话语的时代，上面的故事通常是这样被描述的：库克舰长的"奋进号"军舰对殖民地和尚未成为殖民地的那些地方的所谓"访问"，其实是殖民者耀武扬威的侵略，搭载着达尔文的"小猎犬号"军舰也是同样行径；班克斯和当地女性的纵欲狂欢，当然是殖民者对土著妇女令人发指的践踏；即使是他采集当地植物标本的"科学考察"，也可以视为殖民者"窃取当地经济情报"的罪恶行为。

后来改革开放，上面那种意识形态话语被抛弃了，但似乎又走向了另一个极端，完全忘记或有意回避殖民者和帝国主义这个层面，只歌颂这些军舰上的科学家的伟大发现和成就，例如达尔文随着"小猎犬号"的航行，早已成为一曲祥和优美的科学颂歌。

其实达尔文也未能免俗，他在远航中也乐意与土著女性打打交道，当然他没有像班克斯那样滥情纵欲。在达尔文为"小猎犬号"远航写的《环球游记》中，我们读到："回程途中我们遇到一群黑人姑娘在聚会，……我们笑着看了很久，还给了她们一些钱，这着实令她们欣喜一番，拿着钱尖声大笑起来，很远还能听到那愉悦的笑声。"

　　有趣的是，在班克斯在塔希提岛纵欲六十多年后，达尔文随着"小猎犬号"也来到了塔希提岛，岛上的土著女性同样引起了达尔文的注意，在《环球游记》中他写道："我对这里妇女的外貌感到有些失望，然而她们却很爱美，把一朵白花或者红花戴在脑后的鬓髻上……"接着他以居高临下的笔调描述了当地女性的几种饰物。

　　用今天的眼光来看，这些在别的民族土地上采集植物动物标本、测量地质水文数据等等的"科学考察"行为，有没有合法性问题？有没有侵犯主权的问题？这些行为得到当地人的同意了吗？当地人知道这些行为的性质和意义吗？他们有知情权吗？……这些问题，在今天的国际交往中，确实都是存在的。

　　也许有人会为这些帝国科学家辩解说：那时当地土著尚在未开化或半开化状态中，他们哪有"国家主权"的意识啊？他们也没有制止帝国科学家的考察活动啊？但是，这样的辩解是无法成立的。

　　姑不论当地土著当时究竟有没有试图制止帝国科学家的"科学考察"行为，现在早已不得而知，只要殖民者没有记录下来，我们通常就无法知道。况且殖民者有军舰有枪炮，土著就是想制止也无能为力。正如法拉所描述的："在几个塔希提人被杀之后，一套行之有效的易货贸易体制建立了起来。"

　　即使土著因为无知而没有制止帝国科学家的"科学考察"行为，这事也很像一个成年人闯进别人的家，难道因为那家只有不懂事的小孩子，闯入者就可以随便打探那家的隐私、拿走那家的东西、甚至将那家的房屋土地据为己有吗？事实上，很多情况下殖民者就是这样干的。所以，所谓的"帝国科学"，其实是有着原罪的。

　　如果沿用上述比喻，现在的局面是，家家户户都不会只有不懂事的孩子了，所以任何外来者要想进行"科学探索"，他也得和这家主人达成共识，得到这家主人的允许才能够进行。即使这种共识的达成依赖于利益的交换，至少也不能单方面强加于人。

博物学在今日中国

博物学在今日中国之复兴，北京大学刘华杰教授提倡之功殊不可没。自刘教授大力提倡之后，各界人士纷纷跟进，仿佛昔日蔡锷在云南起兵反袁之"滇黔首义，薄海同钦，一檄遥传，景从恐后"光景，这当然是和博物学本身特点密切相关的。

无论在西方还是在中国，无论在过去还是在当下，为何博物学在它繁荣时尚的阶段，就会应者云集？深究起来，恐怕和博物学本身的特点有关。博物学没有复杂的理论结构，它的专业训练也相对容易，至少没有天文学、物理学那样的数理"门槛"，所以和一些数理学科相比，博物学可以有更多的自学成才者。这次编印的《西方博物学大系》，卷帙浩繁，蔚为大观，同样说明了这一点。

最后，还有一点明显的差别必须在此处强调指出：用刘华杰教授喜欢的术语来说，《西方博物学大系》所收入的百余种著作，绝大部分属于"一阶"性质的工作，即直接对博物学作出了贡献的著作。事实上，这也是它们被收入《西方博物学大系》的主要理由之一。而在中国国内目前已经相当热的博物学时尚潮流中，绝大部分已经出版的书籍，不是属于"二阶"性质（比如介绍西方的博物学成就），就是文学性的吟风咏月野草闲花。

要寻找中国当代学者在博物学方面的"一阶"著作，如果有之，以笔者之孤陋寡闻，唯有刘华杰教授的《檀岛花事——夏威夷植物日记》三卷，可以当之。这是刘教授在夏威夷群岛实地考察当地植物的成果，不仅属于直接对博物学作出贡献之作，而且至少在形式上将昔日"帝国科学"的逻辑反其道而用之，岂不快哉！

2018 年 6 月 5 日
于上海交通大学
科学史与科学文化研究院

《中国昆虫志》
《印度昆虫志》

出版说明

英国博物学家爱德华·多诺万（Edward Donovan）于1768年生于爱尔兰的科克，曾是伦敦林奈学会会员。在那个博物学家和收藏家纷纷开设个人博物馆的时代，多诺万也于1807年在伦敦设立了博物学研究所，展示数百件动物标本与植物标本。

多诺万本人并不去海外收集标本，而是运用自己良好的社会关系，委托包括约瑟夫·班克斯和詹姆斯·库克在内的诸多探险家为自己采集样本，因而得以积攒起大量罕见样本。对标本的研究帮助他出版了诸多负有盛名的博物学著作，如《不列颠珍稀鸟类志》《不列颠贝类志》等。

他最有名的著作，是《中国昆虫志》（1798）和《印度昆虫志》（1800），均配有大量精美插画，细致入微地解析了这些来自遥远异国的小生灵，后者更是第一本向欧洲人介绍印度昆虫的书籍。多诺万笔下的画作亮色鲜艳，并亲自制作铜版，从绘图到着色全部自己包办。加上有包括驻华使节马葛尔尼这样的特殊关系人士带回的珍稀标本，使得这两本著作一经出版便立于当时同类作品的最前沿。

然而，由于需要购买昂贵的标本，加上拿破仑战争后英国经济萧条，多诺万的个人经济状况在十九世纪初期急剧恶化。1817年，他不得不关闭自己的研究所，将藏品拍卖抵债。由于后来的出版事业也一败涂地，多诺万在1837年去世时竟身无分文，债台高筑。

本次西方博物学大系之影印本，将《中国昆虫志》《印度昆虫志》合订为一，底本均为1842年版。

NATURAL HISTORY

OF THE

INSECTS OF CHINA,

CONTAINING

UPWARDS OF TWO HUNDRED AND TWENTY

FIGURES AND DESCRIPTIONS,

BY

E. DONOVAN, F.L.S. & W.S.

A NEW EDITION,

BROUGHT DOWN TO THE PRESENT STATE OF THE SCIENCE,
WITH SYSTEMATIC CHARACTERS OF EACH SPECIES, SYNONYMS, INDEXES, AND OTHER
ADDITIONAL MATTER,

BY J. O. WESTWOOD,

SECRETARY OF THE ENTOMOLOGICAL SOCIETY OF LONDON, HON. MEM. OF THE LITERARY AND HISTORICAL
SOCIETY OF QUEBEC, AND OF THE NATURAL HISTORY SOCIETIES OF MOSCOW, LILLE, MAURITIUS, ETC.

LONDON:
HENRY G. BOHN, YORK STREET, COVENT GARDEN.
MDCCCXLII.

PREFACE.

IN presenting a new edition of the Epitome of the Insects of China to the entomological public, I have endeavoured to bring it down to the present state of the science. The former edition, like all the writings of Donovan, was arranged in accordance with the system of Linnæus, and bore the date of 1798. At that period the science of Entomology was in its infancy; but in the subsequent forty years the progress which has been made, has indeed been rapid. I have endeavoured to render the specific characters more precise, the nomenclature more correct (giving the priority to the oldest specific name), and the synonyms more numerous. The localities in many instances were incorrectly given in the former edition; and I have added many additional observations, either incorporated in the text or given as foot notes, omitting nothing which appeared in the former at all likely to instruct or interest the reader. Alphabetical and systematic indices have also been introduced. I dare not hope that this edition is faultless: I have endeavoured to render the beautiful figures of Donovan as serviceable as possible, and must trust to the indulgence of the more skilful specific entomologist. One circumstance may be mentioned which will, at all events, be deemed an improvement, namely, the introduction of numbers both for the plates and for the several figures on each plate. Those who have consulted synonymical authorities in Entomology are aware of the trouble and confusion which have originated in the want of a

PREFACE.

regular numeration of these plates, which was caused by the original periodical appearance of the work, whence it happens that we are referred to Part II. plate 1. (as it may be), and not to a consecutive series of numbers upon the plates, which were, indeed, entirely without a number, and appeared promiscuously.

A German edition of this work was commenced at Leipzic in 1801, edited by J. G. Grubner, but I am not certain whether the entire work was republished or only one of the parts.

Of the Entomology of China little more is known at the present time than Donovan was acquainted with. It is true we continue to receive numerous boxes of insects from China, chiefly purchased in the shops of Canton, but, like every thing Chinese, there is such an absolute monotony in these arrivals, that it is almost impossible to discover in a quantity of these boxes a single species which is not contained in all the rest. It is evident that a considerable employment is produced by the rearing of the Atlas moth and some other species, and in the collection of the other insects which we receive in such abundance. The Chinese boxes are made of a soft wood, about 16 inches by 11 in size, and of a sufficient depth to admit a tall needle; a layer of butterflies and moths stuck close to the point of the needle is placed at the bottom of the box, with another layer of beetles, flies, &c. closely packed together and stuck high up on the needles, the points of which are passed through the wings of the butterflies forming the under layer.

Donovan well observed that " the Chinese, like their neighbours the Japanese, are well acquainted with the natural productions of their empire, and Zoology and Botany, in particular, are favourite studies amongst them."

That the Chinese also pay considerable attention to Entomology is evident, not only from the fact of the employment of silk having had its origin in that country, but also from the numerous beautiful drawings of insects upon rice

PREFACE.

paper, brought to Europe in great quantities. Many of these figures are, however, evidently fictitious, although some are occasionally found accurately correct and most elaborately pencilled.

The few but interesting hints which Sir George Staunton (who accompanied the embassy of the Earl Macartney to China) has given on the practical Entomology of China in the account of his Travels, which was published shortly before the appearance of the first edition of this work, were embodied by Donovan in his pages, and from whence we may be induced to hope that at some future time, some of the insects, as well as plants, of that vast empire may become no less objects of utility and importance than of curiosity; the Chinese cochineal insect,* and that from which the wax of the East is procured, are two species that deserve particular attention. The medical precepts of the Chinese will certainly find but few votaries in Europe, but as articles of medicine, amongst others, the Mylabris Cichorii, regarded as the Cantharides of the ancients and still used as a vesicant by the Chinese, may be of importance, as it is said to possess more virtues than the Cantharis vesicatoria of Europe.

From the vast extent of the Chinese empire and our comparative ignorance of its insect productions, it is almost impossible to speak with any precision upon the interesting subject of its entomological geography. Many of its insects bear a great resemblance, and occasionally appear identical with those

---

* Dr. Anderson found eight species of Coccus at Madras. One of these, he says, was found on a young citron-tree, citrus sinensis, just landed from China; it was more deeply intersected between the abdominal rings than any of those of the coast, and he therefore named it C. Diacopeis. *Collection of Letters from Madras, January* 28, 1788. The Cactus Cochinillifer had been found previous to the appearance of the first edition by Mr. Kincaid at Canton; its Chinese name is Pau wang. This had been transmitted to the Nopalry of the Hon. East India Company at Madras, and promised to be of future advantage to the commercial concerns of Great Britain.

PREFACE.

of the eastern parts of India. These are, therefore, doubtless the inhabitants of the immediate neighbourhood of Canton, which lies in the same degree of latitude with Bengal. A valuable addition has indeed been made to our knowledge of the entomological productions of Mongolia, and some of the adjacent northern portions of China, by M. Faldermann's publication* of the Coleopterous insects brought from those regions by M. Bungius; from which it would appear that the entomological productions of such portions of China as lie between 40° and 50° of north latitude are quite unlike those of the neighbourhood of Canton lying beneath the tropic of Cancer, bearing, indeed, a far greater resemblance to those of central Russia. It is to be hoped that M. Faldermann will speedily publish the descriptions of the species belonging to the other orders of insects brought from this interesting portion of the globe.

J. O. W.

---

* Coleopterorum ab illustriss. Bongio in China boreali, Mongolia et montibus Altaicis collect. et ab illustr. Turczaninoffio et Sichuckino e provincia Irkutzh missorum. (In Mem. Acad. Imper. des Sciences de St. Petersbourg, Tom. 2. 1835.)

1. Copris Midas.   2. Oryctes Rhinoceros?
3. Oniticellus cinctus.   4. Scarabæus sanctus?
5. Gymnopleurus Sinuatus.

INSECTS OF CHINA.

Order. COLEOPTERA. *Linnæus.*

COPRIS MIDAS.

Plate 1. fig. 1.

Tribe.	Lamellicornes, *Latreille.* (Genus, Scarabæus, *Linnæus.*)
Family.	Scarabæidæ, *Mac Leay.*
Genus.	Copris, *Geoffroy.* (Scarabæus p. *Donovan.*)
Ch. Sp.	C. thorace retuso tricorni, cornu intermedio lato emarginato, capitis clypeo 3-sinuato denteque utrinque laterali valido recurvo armato. Long. Corp. 1½ unc. C. with the thorax blunt in front and armed with three horns, the middle one broad and notched at the tip, the shield of the head with three sinuations, and armed on each side with a thick horn. Length 1½ inch.
Syn.	Scarabæus Midas, *Fabricius Ent. Syst.* I. *p.* 45. *Syst. Eleuth.* I. *p.* 36. *Oliv. Ent.* I. *t.* 30. *f.* 183.

The original description of Fabricius was taken from a specimen in the Banksian cabinet, the locality of America being assigned to it; some mistake must, however, have occurred in this respect, as it is now well known that East India is the true country of this curious species, which has received its specific name, Midas, from the size of the ear-like pair of horns at the sides of the head. The accompanying figure was taken from a specimen in the collection of Drury, which was said to have been received from China. Some interesting observations, on the habits of this species, have been published by Col. Sykes in the first volume of the Transactions of the Entomological Society of London.

ORYCTES RHINOCEROS?

Plate 1. fig. 2.

Family.	Dynastidæ, *Mac Leay.*
Genus.	Oryctes, *Illiger.* Scarabæus p. *Linn.*
Ch. Sp.	O. thorace subbituberculato, capitis cornu simplici, clypeo bifido, elytris punctatis. Long. Corp. 1¾ unc.
	O. with the thorax depressed in front, and with the rudiments of two tubercles, head with a single horn, clypeus bifid, elytra punctate. Length 1 inch, 5 lines.
Syn.	Scarabæus Rhinoceros, *Linn. Syst. Nat.* I. II. p. 544. *Fabricius Syst. Eleuth.* 1. p. 14. *Röesel. Ins.* II. *Scarab.* 1. tab. A. f. 7.
	Scarabæus Nasicornis, *Donovan*, 1st edit.

Donovan observed upon this species, which he considered as identical with the European Oryctes nasicornis, that " the male is furnished with a long recurved horn on the head: the female has only a small rising on that part. It is found in Europe as well as China." It is evident, however, from the form of the thorax, and striation of the elytra of Donovan's figure, that he had confounded two distinct species. I have little doubt that the insect here figured is intended for the Oryctes Rhinoceros, although the elytra are not represented as being punctured.

ONITICELLUS CINCTUS.

Plate 1. fig. 3.

Family.	Scarabæidæ, *Mac Leay.*
Genus.	Oniticellus, *Zeigler.* Scarabæus p. *Fabricius.*
Ch. Sp.	On. niger, elytrorum margine pallido, clypeo emarginato. Long. Corp. lin. 4½.
	On. black with the margin of the elytra pale coloured, the clypeus notched. Length 4½ lines.
Syn.	Scarabæus cinctus, *Fabricius Ent. Syst.* 1. p. 69. *Herbst. Col.* II. p. 327. n. 215. *Oliv. Ent.* 1. 3. 169. 209. t. 10. f. 90.
	Onitis cinctus, *Schonh. Syn. Ins.* 1. p. 33. no. 20.

COLEOPTERA.

SCARABÆUS (HELIOCANTHARUS) SANCTUS?

Plate 1. fig. 4.

Genus.	Scarabæus, *Linn. Mac Leay.* Ateuchus, *Latreille, Dejean.*
(Sub-Gen.	Heliocantharus, *Mac Leay.*)
Ch. Sp.	Sc. cupreo-nitens, clypeo 6-dentato, thorace serrato, elytris striatis. Long. Corp. 1 unc.
	Sc. black slightly shining with brass or copper, clypeus with six teeth, thorax with the margins serrated, elytra striated. Length 1 inch.
Syn.	Copris sanctus? *Fabr. Suppl. Ent. Syst. p.* 34.
	Scarabæus sacer, *Donov. 1st edit.*
	Scarabæus sanctus, *Mac Leay, Horæ Ent. p.* 500, var. δ?

The insect here figured was regarded by Donovan as identical with the Scarabæus sacer of Linnæus, a species inhabiting the south of Europe. "It is," he observes, "a native of China, and is also found in other parts of the East Indies, in Egypt, Barbary, the Cape of Good Hope, and other countries of Africa, and throughout the south of Europe." The insects inhabiting these various countries are now ascertained to be specifically distinct; so that the reference of the species here figured to the Scarabæus sacer cannot be adopted, and it is not improbable that it is identical with the sanctus of Fabricius.

This tribe of insects is especially interesting from its containing the sacred beetle of the Egyptians, by whom it was regarded as a visible deity; but a more refined system of religious worship prevailed in their temples among the priests and sages. They deemed it only the symbol of their god, and, ascribing both sexes to the beetle, it became a striking emblem of a self-created and supreme first cause.*

This insect was more especially the symbol of their god Neith, whose attribute was power supreme in governing the works of creation, and whose glory was increased, rather than diminished, by the presence of a superior being, *Phtha*, the creator. The theological definition of the two powers, being independent, yet centering in one spirit, is implied by the figurative union of two sexes in the beetle. In the latter sense it signified therefore but one omnipotent power. The Scarabæus, typifying Neith, was carved or painted on

---

* "The father, mother, male and female art thou." *Synesius. Hymn. Phtha.*—"The Egyptian spirit *Phtha* gave chaos form, and then created all things." *Jamblichus de Mysteriis, sect.* 8.

COLEOPTERA.

a ring, and worn by the soldiers, as a token of homage to that power who disposed of the fate of battles;* and sculptured on astronomical tables, or on columns,† it expressed the divine wisdom which regulates the universe and enlightens man.

---

* Authors quote a doubtful passage in *Horapollo Hieroglyph. lib.* 1. to support this opinion. That such rings were worn by the ancient Egyptians is beyond conjecture, many remains of them, and some very perfect, have been found in the subterranean caverns and sepulchres in the Plain of Mummies near Saccara and Giza. Those which we have examined are remarkable for the convexity or full *relievo* of the figure sculptured on them; in some it is of the natural size of the insect, but generally smaller; the stone, cornelian, without a rim, and turning on a swivel ring of gold.

† Linnæus says the Scarabœus sacer is sculptured on the antique Egyptian columns in Rome. " Hic in columnis antiquis Romæ exsculptus ab Ægyptiis." *Syst. Nat.* Does Linnæus allude to any remains of those colossal obelisks which Augustus transported to Rome when he subjugated Egypt, or others of more recent date? It would increase the interest of our inquiries to learn that the Scarabæus was among the hieroglyphics, on the two very ancient obelisks, carried from Heliopolis, the city of the Sun.

We are informed by ancient writers, that the Scarabæus engraved on the astronomical tables of these people, implied the divine Wisdom which governed the motion and order of the celestial bodies; that those tables were huge and massy stones or columns of granite, with the characters and figures large and highly embossed; in short, such as were supposed capable of long resistance to the corroding hand of time. Among those the Scarabæus was probably the most conspicuous, its size gigantic, and the figure frequently repeated; for this we have observed even on small Egyptian antiques.

Various valuable remains of tablets, with figures of the *Scarabæus sacer*, are preserved in the British Museum and other collections of antiquities in this country. Those we have examined are of various descriptions, some smaller than the insect itself, others of a monstrous size. The stones on which they are sculptured generally *green nephritic* or *jade stone*, or a kind of *basaltes*, and black marble; the figure *basso relievo* on a tablet or slab, but oftener in *relievo*, with the prominent characters of the insect very accurately defined, particularly the six dentations of the clypeus and those of the tibiæ. The reverse of the embossed side is flat and smooth, and abounds in characters altogether unknown, though, from the number of religious objects of worship occasionally interspersed, we may presume they contain an ample store of the ancient sacerdotal language: the most remarkable were the scarabæus, the sceptre and eye (Osiris), the human figure with a dog's head (Anubis), the hawk (Horus), and the Ibis, or sacred bird. On the thorax of one fine specimen we remarked four elegant figures. One of them is holding a *cornucopia* in the left hand, and a branch in the right: this is perhaps a subordinate deity of the Nile, that river having been once found depictured on an antique Alexandrian coin, like an aged man, holding the cornucopia, and a branch of the Papyrus; denoting its abundance and produce. (In many of the mummies which have been recently unrolled in this country, carved Scarabæi have been found in various positions, especially upon the eye and breast of the body.— See *Pettigrew's History of Mummies.* J. O. W.)

The digression on the mythological history of this insect may be considered by some as a tedious deviation from the pursuit of the naturalist; with others we trust it will be more favourably received; for it proves to the unprejudiced mind how deeply the history of nature, and in the present instance the science of Entomology, involves a most important enquiry into the first philosophical opinions of the human race. The means, however trifling, must not be contemned, which illumine the most sublime of all human researches—*The Study of Mankind.*

1. Copris Molossus. 2. Onthophagus Seniculus.
3. Copris Bucephalus.

COLEOPTERA.

GYMNOPLEURUS SINUATUS.

Plate 1. fig. 5.

GENUS. GYMNOPLEURUS, *Illiger.* Ateuchus p. *Fabricius.*
CH. SP. Gymn. clypeo emarginato, niger subcupreus, antennarum apice flavo. Long. Corp. lin. 8.
G. with the clypeus notched, black with a slight coppery tinge, tips of the antennæ yellow. Length two-thirds of an inch.
SYN. Ateuchus sinuatus, *Fabricius Syst. Eleuth.* 1. p. 60. *Oliv. Ins.* 1. 3. *tab.* 10. *f.* 90. *tab.* 21. *fig.* 189.
Scarabæus Leei, *Donovan, 1st edit.*

The Scarabæus Leei of Fabricius (to which Donovan referred the insect here figured) is totally distinct, and is identical with the Scarabæus fulgidus of Olivier. The original specimen, described by Fabricius, from the collection of the late Mr. Lee, is now in the collection of the Rev. F. W. Hope.

COPRIS MOLOSSUS.

Plate 2. fig. 1.

GENUS. COPRIS, *Geoffroy.* Scarabæus, *Linnæus.*
CH. SP. C. niger, thorace punctatissimo, retuso, bidentato, utrinque impresso; clypeo lunato, ♂ unicorni integro, elytris lævibus. Long. Corp. 1 unc. 4 lin.
C. black, thorax very much punctured, retuse in front and bidentate, with two lateral impressions, clypeus lunate, the male having a single erect horn, elytra smooth. Length $1\frac{1}{3}$ inch.
SYN. Scarabæus Molossus, *Linn. Syst. Nat.* 1. 11. p. 543. *No.* 8. *Fabricius Syst. Eleuth.* 1. p. 42. *Herbst. Col.* 11. p. 178. t. 14. f. 1. *Oliv. Ent.* 3. p. 100. var. c. t. 5. f. 37. ♂. *Drury Ins. Pl.* 32. f. 2. *2nd edit.* p. 64.

S. Molossus and S. Bucephalus are very common in China. The first seems a local species; the latter is said to be found in other parts of the East Indies. Olivier has given three varieties of Scarabæus Molossus. The specimen figured in the annexed plate is the *var. c.* of that author.

The larvæ of the larger kinds of coleopterous insects, abounding in unctuous moisture, are not less esteemed as food among some modern nations, than they were by the epicures of antiquity. In Jamaica and other islands in the West Indies, the larva of the Prionus damicornis, or Macokko beetle, is an article of luxurious food; and in China many insects in that state are appropriated to the same purpose. Thus, also, the Romans introduced the larvæ of the Lucani and Cerambyes in their voluptuous repasts; previously feeding them on farinaceous substances to give consistence to the animal juices.

The learned author of the last account we have of China, says, "Under the roots of the canes is found a large white grub, which being fried in oil is eaten as a dainty by the Chinese." Donovan suggests that perhaps this is the larva of Scarabæus Molossus, which, like many other of the Scarabæi,* may live sedentary in the ground, and subsist on the roots of plants: the general description and abundance of this insect in China favours such opinion. The same author observes in another part of his work, that "the aurelias of the silk worm which is cultivated in China, after the silk is wound off, furnish an article for the table." This also is a very ancient custom among the Asiatics, and even Europeans before the sixteenth century, if we may credit Aldrovandus:† it is certain the worms, if not the chrysalides, were administered in medicine in early ages.‡ Fabricius also expressly states that the insect here figured is medicinally employed in China.

ONTHOPHAGUS SENICULUS.

Plate 2. fig. 2ª. and 2ᵇ.

GENUS.	ONTHOPHAGUS, *Latreille.* Scarabæus p. *Fabricius, Donovan.*
CH. SP.	O. thorace antice, clypeo postice bicorni, elytris substriatis strigis duabus baseos e punctis ferrugineis, punctoque uno alterove flavescentis apicis. Long. Corp. lin. 5½.
	O. with the thorax in front, and the hinder part of the clypeus with two horns, elytra slightly striated, with two rows of basal pale spots and with two yellowish apical spots. Length nearly half an inch.
SYN.	Scarabæus seniculus, *Fab. Ent. Syst.* 1. p. 43. 142. *Oliv. Ent.* 1. 3. p. 124. t. 7. f. 56. a. b Panzer in *Naturforscher*, 24. t. 1. f. 5.
	♀ Scarabæus brevipes, *Herbst. Archiv.* t. 19. f. 16.
HABITAT.	Tranquebar (*Fabricius*), China, *Donovan, Weber MSS.*

The annexed figures exhibit the two sexes of Scarabæus Seniculus. In some specimens the spots are very indistinct and reddish; in others the wing-cases have faint red striæ. The female has the rudiments of horns on the thorax.

---

* The larvæ of the Scarabæi live in the trunks of decayed trees, in putrid and filthy animal substances, or in the earth. The last are the most injurious, because they destroy the roots of plants. All the known kinds of these larvæ are of an unwieldy form and whitish colour, the skin free from hairs, and only the head and fore feet defended with a shelly covering. (As it is most probable that the habits of this large Copris are analogous to those of the English C. lunaris, I should be rather inclined to regard the cane grub mentioned in the above extract as the larva of a Calandra. J. O. W.)

† The German soldiers sometimes fry and eat silk worms.   *Aldrov.*

‡ Silk worms dried, powdered, and put on the crown of the head, help the *vertigo* and *convulsions*; mundify or cleanse the blood, &c. &c.   *Schroderus, Serapio, &c. &c.*

1. Cetonia Chinensis. 2. Euchlora viridis.

COLEOPTERA.

COPRIS BUCEPHALUS.

Plate 2. fig. 3.

GENUS. COPRIS, *Geoffroy*. Scarabæus p. *Linn. &c.*
CH. SP. C. thorace retuso quadridentato, capitis clypeo angulato, cornu ♂ brevi erecto haud emarginato, ♀ breviori truncato subemarginato. Long. Corp. unc. 1, lin. 8.
C. with the thorax retuse in front and 4-toothed, clypeus angulated, the male with a short erect horn not notched at the tip, the female with a much shorter and truncated horn. Length 1¾ inch.
SYN. Scarabæus Bucephalus, *Fabricius Ent. Syst.* 1. *p.* 51. 166. *Herbst. Col.* II. *p.* 174. *t.* 13. *f.* 1. 2. *Oliv.* I. 3. 99. *t.* 4. *f.* 26. *t.* 10. *f.* 92. *t.* 22. *f.* 92. ♀

This species has been confounded with C. Molossus. Both species are here represented on the same plate, in order that the difference between them may be precisely observed.

CETONIA (TETRAGONA) CHINENSIS.

Plate 3. fig. 1 and 1ª.

FAMILY. CETONIIDÆ.
GENUS. CETONIA, *Fabricius.* Scarabæus p. *Linnæus, Donovan.*
SUB-GEN. TETRAGONA, *Gory.*
CH. SP. C. viridi-ænea, clypeo emarginato, thorace posticè lobato, elytris acuminatis, corpore subtus pedibusque subspinosis, castaneis, tarsis nigris. Long. Corp. unc. 1, lin. 6.
C. brassy green, clypeus notched and slightly spined, thorax lobed behind, elytra acuminated, body beneath, with the legs, chestnut-coloured, tarsi black. Length 1½ inch.
SYN. Scarabæus Chinensis, *Forster Cent. Ins.* I. *p.* 2. 2.
Cetonia Chinensis, *Fabr. Ent. Syst.* I. II. *p.* 126. *b.* *Oliv. Ent.* I. 6. *p.* 11. *t.* 2. *f.* 5. *a. b.* *Herbst. Col.* III. *p.* 199. 2. *t.* 28. *f.* 2.
Scarabæus oblongus, *Brown Illustr. t.* 49. *f.* 4.
Smaragdinus major, *Voet. Col. t.* 5. *f.* 40.

EUCHLORA VIRIDIS.

Plate 3. fig. 2.

FAMILY. MELOLONTHIDÆ, *Mac Leay.*

GENUS. EUCHLORA, *Mac Leay.* Scarabæus p. *Linnæus, &c.*

CH. SP. E. supra viridis subtilissime punctata, subtus cum pedibus aurea vel cupreo-ænea. Long. Corp. lin. 11.

E. green above, and very finely punctured, body beneath with the legs golden or coppery bronze coloured. Length nearly 1 inch.

SYN. Melolontha viridis, *Fabricius Ent. Syst.* I. II. p. 160. *no.* 23. *Oliv. Ent.* 1. 5. p. 29. 31. t. 3. f. 21. *Herbst. Col.* III. p. 149, t. 26. f. 5. *Mac Leay, Horæ Entomol.* 1. p. 148. (Euchlora v.)

RHINASTUS STERNICORNIS.

Plate 4. fig. 1.

TRIBE. RHYNCOPHORA, *Latreille.*

FAMILY. CURCULIONIDÆ, *Leach.*

DIVISION. CHOLIDES, *Schonherr.*

GENUS. RHINASTUS, *Schonherr.* Cholus p. *Germar.* Curculio p. *Donovan.*

CH. SP. Rh. longirostris, femoribus dentatis, corpore polline flavescente obtecto lateribus nigris, rostro utrinque spinoso. Long. Corp. (sine rostro) lin. 11.

Rh. with the rostrum long, with a short spine on each side at the tip, thighs toothed, body clothed with a fine yellow powder, except a black stripe on the sides of the thorax and elytra. Length (without the rostrum) nearly 1 inch.

SYN. Rhinastus sternicornis, *Schonherr. Syst. Curcul.* vol. 3. p. 558.
Cholus sternicornis, *Germar. Ins. spec.* 1. p. 214. t. 1. f. 4.
Curculio Chinensis, *Donovan,* 1st edit.

This insect seems nearly allied to *Curculio mucoreus,* an Indian species, described by Linnæus, but not figured by any author: the lateral stripe of black, and the denticulations on the posterior thighs of our insect clearly remove it, however, from the Linnæan species.

Except the lateral black stripes, and the rostrum, this insect is totally covered with a bright brown powder, or rather, with very minute hairs which adhere but slightly, and resemble that substance. We observe a similar farinaceous appearance on the *Curculiones, Lacteus, Niveus, &c.* and especially on that gigantic beetle *Scarabæus Elephas.*

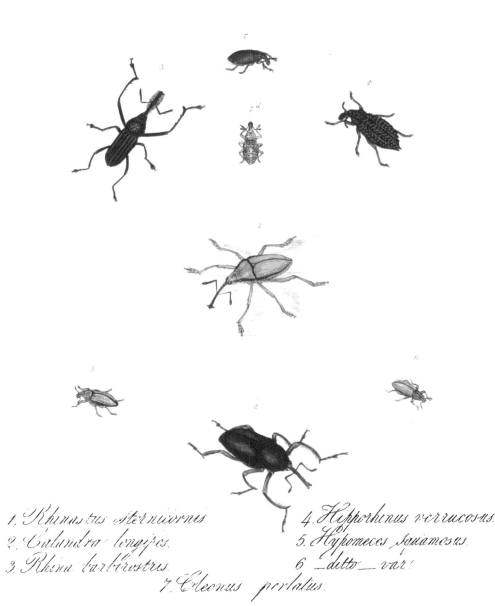

1. Rhinastus sternicornis.
2. Calandra longipes.
3. Rhina barbirostris.
4. Hipporhinus verrucosus.
5. Hypomeces squamosus.
6. ditto var.
7. Cleonus perlatus.

I have not the least doubt that this insect is identical with the Brazilian species, Rhinastus sternicornis, of Schonherr, although the specific name given to it by Donovan implies it to be a native of China. Some mistake had doubtless arisen relative to the original country of which it is an inhabitant which led our author into this error

CALANDRA LONGIPES.

Plate 4. fig. 2.

DIVISION. RHYNCOPHORIDES, *Schonherr.*
GENUS. CALANDRA, *Fabricius.* Rhyncophorus, *Herbst.* Curculio p. *Linn, &c.*
CH. SP. C. nigricans; elytris ferrugineis, rostro apice utrinque reflexo (in uno sexu dorso bi serrato,) pedibus anticis elongatis. Long. Corp. (excl. rostro lin. 13.)
C. blackish-brown, elytra ferruginous; rostrum, with the apex on each side reflexed, and in one sex with the upper edge doubly serrated, the fore legs long. Length 13 lines.
SYN. Curculio longipes, *Fabricius, Ent. Syst.* 2. 395. *Syst. Eleuth.* 2. 431. *Oliv. Ins.* 83. *tab.* 86. *fig.* 191. (See *Drury, Illustr. vol.* 2. *t.* 33.)

In his earlier works, Fabricius erroneously gave the Cape of Good Hope as the locality for this very common Chinese species.

RHINA BARBIROSTRIS.

Plate 4. fig. 3.

GENUS. RHINA, *Latreille.* Curculio p. *Donovan, &c.*
CH. SP. Rh. longirostris; niger, rostro barbato, tibiis anticis tridentatis. Long. Corp. (rostr. excl.) lin. 11.
Rh. with a long rostrum, its sides thickly clothed with hairs on the male, black, punctured, anterior tibiæ long and armed with three spines beneath. Length without the rostrum 11 lines.
SYN. Curculio barbirostris, *Fabricius Ent. Syst.* 1. *p.* 2. 418.
Rhina barbirostris, *Latreille Genera, vol.* 2. *p.* 269.

Donovan, misled by the Fabrician reference of this insect to the locality " In Indiis," introduced it into the present work. It is, however, a native of South America. Drury has figured the female of another species in his Illustrations, vol. 2. pl. 34. f. 2. from the Island of Johanna.

HIPPORHINUS VERRUCOSUS.

Plate 4. fig. 4.

DIVISION. ENTIMIDES, *Schonherr.*
GENUS. HIPPORHINUS, *Schonh.* Hipporhis, *Bilberg.* Curculio p. *Auct.*
CH. SP. H. elongato-ovatus, niger, æneo-micans, thorace confertim tuberculato, elytris seriato-tuberculatis apice singulatim verruca crassa auctis. Long. Corp. lin. 9.
H. elongate-ovate, of a black colour, slightly tinged with brass, thorax closely tubercled, elytra with three rows of elevated warts on each, with intermediate double rows of punctures, and a large tubercle at the tip of each. Length ¾ inch.
SYN. Curculio verrucosus, *Linn. Syst. Nat.* I. II. *p.* 618. *Fabricius Syst. Eleuth.* 2. 534. *Schonh. Syn. Ins. Curcul.* 1. *p.* 481. *Drury Illustr.* 1. pt. 32. *f.* 5.

Donovan correctly referred this insect to the Curculio verrucosus, but incorrectly gave it as an inhabitant of China. This species, as well as the numerous insects of which the genus Hipporhinus is composed, are inhabitants of the Cape of Good Hope or New Holland.

HYPOMECES SQUAMOSUS.

Plate 4. fig. 5. magn. nat.
Plate 5. idem magn. auct.
Plate 4. fig. 6. idem, var. β.

DIVISION. BRACHYDERIDES, *Schonherr.*
GENUS. HYPOMECES, *Sch.* Curculio p. *Auct.*
CH. SP. H. oblongus, niger, squamulis viridi-aureis undique tectus, thorace supra fere plano, ruguloso, longitudinaliter canaliculato elytris subtiliter punctato striatis. Long. Corp. lin. 7.
H. oblong black, entirely covered with golden green scales, thorax nearly flat above, and rugose with a longitudinal canal, elytra slightly punctate-striate. Length 7 lines.
VAR. β. Corpore squamulis cinerascentibus undique tecto. Entirely clothed with grey scales.
SYN. Curculio squamosus, *Fabricius Ent. Syst.* 1. *p.* 2. 452. *Syst. Eleuth.* 2. 510. *Oliv. Ent.* 5. *p.* 319. *t.* 5. *f.* 48. *a. b. Schonherr Syn. Ins. Curcul.* 2. *p.* 71.
VAR. β. *Schonherr* loc cit. Curculio pulverulentus, *Fabricius Syst. Eleuth.* 2. 510. *Donovan*, 1st. edit.
Curculio orientalis, *Oliv. Ent. v. p.* 321. *t.* 6. *f.* 66?

A small, but superb species, being totally covered with minute scales of an oblong form, and resplendent green colour, interspersed with changeable sparks of gold and

Pl. 5

Hyphomeces Squamosus

1. Lamia Rubus. 2. Lamia Reticulator.
3. Lamia punctator.

crimson, in various reflections of light. This scaly covering is not unlike that of Polydrusus argentatus found in England; but of a brilliance scarcely inferior to the gem-like spangles on the Entimus imperialis of Brazil.—Hypomeces squamosus is represented in its natural size, in the annexed plate; and, in justice to an insect of such uncommon beauty, an additional plate is given, to exhibit its appearance in the opaque microscope. —It is extremely common in China.

The insect represented in fig. 6 is entirely black when divested of its scaly covering. Fabricius considered it as specifically distinct from H. squamosus, which opinion was adopted by Donovan. It is, however, considered by Schonherr, and other recent writers, as a variety only of that species, the C. unicolor, Fabr. and C. rusticus, Weber, being also equally regarded as varieties.

CLEONIS PERLATUS.

Plate 4. fig. 7.

DIVISION. CLEONIDES, *Schonherr*.

GENUS. CLEONIS, *Schonherr*. Cleonus, *Dejean*. Curculio p. *Linn. &c.*

CH. SP. Cleonis niger, subtus densè, supra tenuiter cinereo-albido tomentosus, thorace vittis sub-tribus albidis notato, elytris punctato-striatis, ventre tuberculis numerosis glabris atris notato. Long. Corp. lin. 6.

C. black; beneath densely, above slightly clothed with greyish scales, thorax with three slight whitish stripes; elytra punctate-striate, abdomen marked with numerous black smooth tubercles. Length ½ inch.

SYN. Curculio perlatus, *Fabricius Syst. El.* II. p. 516.
Lixus faunus, *Oliv. Ent.* V. p. 267. *t.* 24. *f.* 342.

LAMIA RUBUS.

Plate 6. fig. 1.

TRIBE. LONGICORNES, *Latr.* (Cerambyx *Linn.*)

FAMILY. LAMIIDÆ.

GENUS. LAMIA, *Fabricius*. Cerambyx p. *Linn, &c.*

CH. SP. L. grisea, thorace utrinque spinoso bimaculato, elytris basi scabris, humeris apiceque mucronatis, albo-guttatis. Long. Corp. unc. 1. lin. 9.

L. grey coloured, thorax on each side with a spine and with two oblong spots on the back, elytra rough at the base, with the shoulders and apex spined, spotted with white. Length 1¾ inch.

SYN. Cerambyx Rubus, *Linn. Syst. Nat.* 1. II. 625. 21.
Lamia Rubus, *Fabricius Ent. Syst.* 2. 290. *Oliv. Ent.* 67. *t.* 7. *f.* 57.

This is the largest species of this genus found in China. It is also very abundant in different parts of the East Indies. Some interesting observations upon its habits have been published by W. W. Saunders, Esq. F. L. S. in the first part of the Transactions of the Entomological Society of London.

LAMIA RETICULATOR.

Plate 6. fig. 2.

Ch. Sp.   L. nigra, thorace elytrisque fulvis, thorace nigro-lineato, elytris reticulatis, antennarum articulo 3^{tio} fasciculato. Long. Corp. lin. 13.

L. black, thorax and elytra rich golden brown, the former marked with black lines, and the latter with irregular black marks, antennæ with the third joint furnished with a whorl of hairs. Length 13 lines.

Syn.   Lamia reticulator, *Fabricius Ent. Syst.* 1. *p.* 2. 278. *no.* 44.   *Oliv. Ins.* 67. *tab.* 12. *f.* 85.

This is altogether a beautiful insect; but the singular structure of the antennæ deserves particular notice: it is entirely brown except the first articulation, which is black; the third has a large verticillated tuft of black hair at the summit; at the base of this articulation it has another tuft, but smaller; and a similar tuft, but still smaller, is situated on the two following articulations.

LAMIA PUNCTATOR.

Plate 6. fig. 3.

Ch. Sp.   L. atra, elytris albo punctatis, antennis longis articulis albo nigroque variis. Long. Corp. 1 unc. 3 lin.

L. black shining, the elytra with farinaceous white spots, the antennæ long, the joints varied with black and white. Length $1\frac{1}{4}$ inch.

Syn.   Lamia punctator, *Fabr. Sp. Ins.* 1. 221. 30.   *Syst. Eleuth.* 2. 298.   *Oliv. Ent.* 4. 69. *t.* 8. *f.* 50. *a. b.*   *Schonh. Syn. Ins.* 3. *p.* 386.
Cerambyx Chinensis, *Forster Cent. Ins.* 39.
Cimex farinosus, *Drury, Edit.* 1ma *vol.* 2. *pl.* 31. *f.* 4.   *Donovan, Edit.* 1^{ina} *(nec Linn. Syst. Nat.* 1. 2. 626.)

Donovan states that among some Chinese drawings of the late Mr. Bradshaw, he observed one on which the metamorphosis of this insect was delineated. The larva was partly concealed in the hollow of a piece of decayed wood; it was of a whitish colour, with the head and tail black, as described by Fabricius. The true Cerambyx farinosus of Linnæus, with which this insect was confounded by Donovan, is an inhabitant of South America.

1. Buprestis Vittata. 2. Buprestis Ocellata.

COLEOPTERA.

BUPRESTIS (CHRYSOCHROA) VITTATA.

Plate 7. fig. 1.

TRIBE.	STERNOXI, *Latreille.*
GENUS.	BUPRESTIS, *Linnæus.*
SUB-GEN.	CHRYSOCHROA, *Carcel et Delap.*
CH. SP.	B. aureo-viridis, elytris bidentatis, punctatis; lineis quatuor elevatis, vittaque lata aurea. Long. Corp. 1 unc. 5 lin.
	B. golden green, elytra with two apical teeth, punctured, with four elevated lines and a broad golden stripe. Length 1 inch 5 lines.
SYN.	Buprestis Vittata, *Fabricius Syst. Eleuth.* 2. *p.* 187. *Oliv. Ent.* 2. 32. *p.* 9. *t.* 3. *f.* 17. *a—d. Herbst. Col.* 9. *tab.* 138. *f.* 4.
	Buprestis ignita, *Herbst. Archiv. t.* 28. *a. f.* 3.

The family Buprestidæ is one of the most extensive and brilliant tribes of coleopterous insects. Brasil and New Holland produce some gigantic species, but none more beautiful than those of India. We need adduce no other proof of this, than Buprestis chrysis, sternicornis, attenuata, ocellata, and vittata. These wrought into various devices and trinkets decorate the dresses of the natives in many parts of India. The Buprestis vittata in particular is much admired among them. It is, we believe, entirely peculiar to China, where it is found in vast abundance, and distributed from thence at a low price among the other Indians. The Chinese, who always profit by the curiosity of Europeans, collect vast quantities of this Buprestis, and other gay insects, in the interior of the country, and traffic with them.

The Buprestis ignita of Linnæus, with which the present species has been partially confused, has not the brilliance of colours that so eminently distinguishes B. vittata, but in form and size it agrees with it. The only figure of that species is given by Olivier, from a specimen formerly in the cabinet of Gigot d'Orcy, of Paris. The specimen in the cabinet of Sir J. Banks, referred to by Fabricius as B. vittata, agrees with Sulzer's figure of that species, as well as the specimen represented here, so that the reference by Fabricius of Sulzer's figure to B. ignita is incorrect.

Fabricius has given as a part of the specific distinction of these insects, that B. ignita has *three spines* at the end of each wing case, or elytron, and B. vittata no more than *two*. This may form a sufficient characteristic in those species; but we must remark, that it is not so in Buprestis ocellata. We have two specimens that have two spines at the end of each elytron, and another with three, as Fabricius has described it. We also find several insects nearly allied to B. vittata, the stripe of gold on each side excepted; one of these has six teeth, another four teeth, and a third only two.

Donovan observes that the Buprestides are supposed, for the most part, to undergo their

COLEOPTERA.

transformations in the water, or marshy ground. This opinion cannot, however, be adopted, as it is now well ascertained that they reside in the early stages of their existence in timber. The Chinese plant represented is the Canna Indica, or Indian flowering Reed.

BUPRESTIS (CHRYSOCHROA) OCELLATA.

Plate 7. fig. 2. upper, and 2ª under side.

Ch. Sp.   B. viridi-nitens, elytris lineis tribus elevatis, macula ad basin alteraque apicali aureis, ocello magno flavo. Long. Corp. 1 unc. 4 lin.

B. shining-green, elytra with three elevated lines, a large yellow round spot in the middle of each, having a golden red spot above, and another behind it. Length 1¼ inch.

Syn.   Buprestis Ocellata, *Fabricius Syst, Eleuth.* 2. *p.* 193. *no.* 38. *Oliv. Ent.* 2. *p.* 27. *t.* 1. *f.* 3. *a. b. Herbst. Col.* IX. *p.* 70. *t.* 144. *f.* 1. *De Geer Ins.* 7. 633. *tab.* 47. *f.* 12.

The Buprestis ocellata is very rare. Olivier says it is from *Chandernagore* in the East Indies. Mr. Drury possessed an extraordinary variety of this insect from China, in which the two spots united at the suture so as to form only one large spot on the back when the wing cases are closed.

These spots are strikingly characteristic of this species. They are situated in the centre of each elytron; are somewhat pellucid, and in fine specimens are cream colour, surrounded with a crimson circle. These spots are sometimes brown; probably they become so after the insect dies.

Our figures represent this insect with expanded wings; one of those is designed to exhibit the beautiful appearance of the under surface, particularly the effulgent abdomen and purple colour of the interior part of the elytra.

MYLABRIS CICHORII.

Plate 8. fig. 1. and 1ª.

Tribe.   Trachelides, *Latreille.*
Family.   Meloidæ.
Genus.   Mylabris, *Fabricius.* Meloe p. *Linn.*
Ch. Sp.   M. nigra elytris maculâ basali fasciisque duabus latis undatis fulvis. Long. Corp. 9—12 lines.

M. black, each of the elytra with a round basal spot, and two broad irregular bands of a fulvous colour. Length from ⅞ths to 1 inch.

Syn.   Meloe cichorii, *Linn. Syst. Nat.* 1. II. *p.* 680. 5. *Fabr. Syst. Eleuth.* 11. *p.* 81. 2. *Schonh. Syn. Ins. vol.* 3. *p.* 31. *Billb. Monagr. Mylabr. p.* 11. 4. *t.* 1. *f.* 8.

1. *Mylabris Cichorii.*   2. *Sagra Splendida.*

This insect is very common in China and some other parts of the East Indies. The small specimen (fig. 1. a.) is rare, and is, probably, the male. According to Olivier, " the Cantharides of the ancients, and those of the Chinese, are not the same as ours. The Chinese employ the *Mylabris Cichorii*, and it appears from *Dioscorides Mat. Med. Lib. 2. Cap.* 65, that the ancient Cantharides were the same as those now used by the Chinese." " The most efficacious sort of Cantharides," says Dioscorides, " are of many colours, having yellow transverse bands; the body oblong, big, and fat; those of only one colour are without strength." The description Dioscorides has given does not agree with our species of Cantharides, as they are of a fine green colour, but is more applicable to the *Mylabre de la Cichorei*, which is very common in the country where Dioscorides lived. *Olivier's Entomologie, ou Hist. Nat. des Insectes. Vol. I. Introd.*

By the term Cantharides, in an European *Pharmacopœia*, we understand the Meloe vesicatorius* of Linnæus, an insect whose medicinal properties are very generally known.† The Cantharides of the ancients can scarcely be ascertained; it was a term indiscriminately applied to several kinds of insects, and too often without regard to their physical virtues. Pliny speaks of the Cantharis as a small beetle that eats and consumes corn; and of another that breeds in the tops of ashes and wild olives, and shines like gold. The ancients were certainly well acquainted with our common sort, though it is confounded with others in a general appellation.‡ *Hippocrates, Galen, Pliny, Matthiolus,* and other physical writers of antiquity, treat of the medicinal uses of Cantharides; but it is not clear that they alluded to only one species : indeed Dioscorides also mentions those of only one colour as being employed as vesicants. The ancients often confounded the term Scarabæus with Cantharis; but whether because they knew that the common kinds of Scarabæi produce the same effects as the Cantharis, is uncertain.—The *Scarabæus auratus*, and *Melolontha*, several *Coccinellæ, Cimex nigro-lineatus, &c. &c.* have a place in the *Materia Medica* as *Cantharides*.

---

* Geoffroy calls this a Cantharis. The Linnæan Cantharis is a distinct genus. (Telephorus, *Latreille*.)
† Applied externally to raise blisters. It is a violent poison taken inwardly, except in small portions.
‡ The common sort has been called *Musca Hispanica* by some Latin authors, and hence Spanish fly by Boyle.

COLEOPTERA.

SAGRA SPLENDIDA.
Plate 8. fig. 2. ♂ fig. 2ᵃ. ♀?

TRIBE.	EUPODA, *Latreille.*
FAMILY.	SAGRIDÆ.
GENUS.	SAGRA, *Fabricius.* Tenebrio p. *Linn, &c.*
CH. SP.	S. cyaneo-purpurea, femoribus posticis dentatis, tibiis apice sinuatis. Long. Corp. lin. 9.
	S. glowing purple, changeable to yellow or green, posterior femora dentate, tibiæ with a deep notch at the tips. Length 9 lines.
SYN.	Sagra splendida, *Weber Obs. Ent. p.* 61. *Fabricius Syst. Eleuth.* 2. *p.* 27.
	Sagra femorata, *Donovan,* 1st edit. in tab.
	Tenebrio femoratus, *Donovan.* 1st edit. in text.

Donovan correctly observed, that the insect here figured differed from the figures referred to by Fabricius under Sagra femorata, the only described species at the period of the publication of the first edition of this work. Weber, however, subsequently published a monograph upon the genus, and this Chinese species appears to be identical with his Sagra splendida. The smaller figure appears also to be identical with the Sagra purpurea of Weber, which Fabricius was inclined to regard as the male of the former, although he gave it as distinct with the character:—" Sagra purpurea, nitida, femoribus posticis unidentatis tibiis integris." It appears to me, however, that it is merely the female of S. splendida.

Order. ORTHOPTERA. *Olivier.*

MANTIS (SCHIZOCEPHALA) BICORNIS.
Plate 9. fig. 1.

TRIBE.	CURSORIA, *Latreille.*
FAMILY.	MANTIDÆ.
GENUS.	MANTIS, *Linn.*
SUB-GEN.	SCHIZOCEPHALA, *Serville.*
CH. SP.	M. thorace filiformi lævi, testaceo, oculis oblongis porrectis acuminato spinosis, elytris alis brevioribus. Long. Corp. 4¼ unc.
	M. with the thorax filiform smooth and pale buffish coloured, eyes oblong, porrected, and produced into a sharp point, elytra shorter than the wings. Length 4¼ inches.
SYN.	Mantis bicornis, *Linn. Syst. Nat.* I. 11, *p.* 691.
	Mantis oculata, *Fabricius Ent. Syst. t.* 2. *p.* 19.
	Schizocephala stricta, *Serville Revis. Orthopt. p.* 29 ?
	La Mante Chinoise étroite cornue. *Stoll. Repres. des Mantes.*

1. Mantis bicornis. 2. Empusa Flabellicornis.

ORTHOPTERA.

Two figures of insects, very much resembling our species, are given in the work of Stoll, &c.; one kind he calls *La Mante étroitement cornue,* the other *La Mante Chinoise étroite cornue.* The first is from the coast of Coromandel and Tranquebar, the other, as its name implies, is a Chinese insect. Donovan states that he could not discover any material difference between these figures and the specimen here figured, and was inclined to consider them altogether as one species.

It is a considerable disadvantage to the works of Stoll, as well as to the naturalist who consults them, that no scientific names, or definitions, are given to the figures of many rare insects included amongst them, hence they have been but rarely referred to by Fabricius.

The Mantis oculata of Fabricius is an African insect, and was described from the collection of the Right Hon. Sir J. Banks, Bart., Donovan compared his Chinese specimen with it, and found it precisely the same species.

MANTIS (EMPUSA) FLABELLICORNIS.

Plate 9. fig. 2.

Sub-Gen. Empusa, *Latreille.* Gongylus, *Thunberg.*

Ch. Sp. M. thoracis parte anticâ dilatatâ membranaceâ; femoribus anticis spinâ, reliquis lobo terminatis, antennis pectinatis. Long. Corp. 2½ unc.

M. with the front of the thorax furnished with a large membrane, the two anterior thighs terminated by a spine, and the four posterior by a rounded membranous lobe, antennæ pectinated. Length 2½ inches.

Syn. Mantis flabellicornis, *Fabricius, Ent. Syst.* II. p. 16. no. 16. *Serville Revis. Orthopt.* p. 21. (Empusa fl.)

This Mantis is described by Fabricius only. Stoll has given the figure of an insect not unlike it in his publication; and we have seen a specimen similar to it, which was found by Professor Pallas near the Caspian sea. It is allied to Mantis Gongylodes,[*] a native of Africa and Asia, but bears a closer affinity to Mantis Pauperata[†] from Java, Molucca, and perhaps other islands in the Indian sea.

Fabricius enumerates fifty-one species of this genus in his last system; a considerable portion of these are from Asia: had he included the America and New Holland species, his genus would have been far more comprehensive. Few naturalists have had the opportunity of observing the manners of these creatures in distant countries; nor can we always rely on the information those few have given. Of the European species we can

---

[*] Serville, indeed, considers it possible that M. flabellicornis may be the male of M. gongylodes.
[†] Figured by Stoll under the name of *La Mante Goutteuse Brune?*

D

ORTHOPTERA.

speak with more precision, because some indefatigable naturalists have attended minutely to them; Roesel in particular has treated at considerable length on the manners of the Mantis religiosa of Linnæus.

Descriptions can only convey an imperfect idea of the extraordinary appearance of many creatures included in the genera Mantis and Locusta. Among them are found species that bear a similitude to the usual forms of other insects; but, from these we almost imperceptibly descend to others, bearing as strong a similitude to the vegetable part of creation; seeming as if Nature designed them to unite the appearance of a vegetable with the vital functions of an animal, to preserve them from the ravages of voracious creatures, or to connect that chain of progressive and universal being, which

" The great directing MIND of ALL ordains."

Many of these creatures assume so exactly the appearance of the leaves of different trees, that they furnish the entomologist with unerring specific distinctions; thus we have L. *citrifolia*, *laurifolia*, *myrtifolia*, *oleifolia*, *graminifolia*, and others, equally expressive of their resemblance in form, and colours, to the leaves of those respective plants. Travellers, in countries that produce these creatures, have been struck with the phenomenon, as it must appear, of animated vegetable substances; for the manners of the Mantis, in addition to its structure, are very likely to impose on the senses of the uninformed. They often remain on the trees for hours without motion, then suddenly spring into the air, and, when they settle, again appear lifeless. These are only stratagems to deceive the more cautious insects which they feed upon; but some travellers who have observed them, have declared they saw the leaves of those trees become living creatures, and take flight.

M. Merian informs us of a similar opinion among the Indians, who believed these insects grew like leaves on the trees, and when they were mature, loosened themselves and crawled, or flew away. But we find in the more pretending works of Piso similar absurdities.

" Those little animals," says that author, " change into a green and tender plant, which is of two hands breadth. The feet are fixed into the ground first; from these, when necessary humidity is attracted, roots grow out, and strike into the ground; thus they change by degrees, and in a short time become a perfect plant. Sometimes only the lower part takes the nature and form of a plant, while the upper part remains as before, living and moveable: after some time the animal is gradually converted into a plant. In this Nature seems to operate in a circle, by a continual retrograde motion." *

---

* Donovan quoted, in a note, Ovid's account of the Transformation of Phaeton's Sisters into trees. " Luna quater junctis implerat cornibus orbem," &c. which he seems to think had its origin in some such idea as this.

ORTHOPTERA.

Roesel treats this account with more than merited severity; not because he could contradict the relation of Piso, but, because he had never observed the same circumstance attend the Wandering Leaf, or Mantis Oratoria, in Europe;* although he afterwards describes even the first symptom of the transformation as related by Piso. When he speaks of the death of the European species, his words are, " As their dissolution approaches, their green eyes become brown, and they unavoidably lose their sight: they remain a long while on the same spot, till at last they fall quite exhausted and powerless, as if asleep." As to the change after they remained long on the ground, such as sending forth fibres, roots, and stems, from the body of the insect, it is only astonishing such a well-informed naturalist should have deemed it matter of surprise. Could he be ignorant of the many instances that occur, of animal substances producing plants?† or was he not informed that the pupa which commonly sends forth a bee, a wasp, or cicada, has sometimes become the nidus of a plant, thrown up stems from the fore part of the head, and changed in every respect into a vegetable, though still retaining the shell and exterior appearance of the parent insect at the root?‡ We own at first sight with Roesel that the account of Piso seems " an inattentive and confounded observation," but that an insect may strike root into the earth, and, from the co-operation of heat and moisture, congenial to vegetation, produce a plant of the *cryptogamic* kind, cannot be disputed. We have seen species of *clavaria* both of the undivided and branched kinds, four times larger than the insect from which they sprang; and can we then deny that the insect mentioned by Piso might not produce a plant of a proportionate magnitude? In short, we are not sufficiently acquainted with the productions of Brazil to contradict any of his assertions, concerning this transformation. Piso does not say of what kind this vegetable was; it must surely be of the fungi kind: reasoning then from analogy, it might be an unknown species of *clavaria* with numerous and spreading branches; and, finally, the colour of

---

* Among the annotations on the last edition of Roesel's *Insecten Belustigung* we find one relating to this part of the works of Piso. " *Der seel Her geheime Rath Trew, &c.* Couns. Trew assures Mr. Roesel that *Piso* not only very often gave out the credible observations of others as his own, but himself believed the most incredible relations, and pretended to be an eye witness thereof." We quote this in justice to the remarks of Roesel. *Note in page* 10, *section Das Wandlende Blat.*

† Such as Mucor crustaceus, &c.

‡ Specimens of these vegetated animals are frequently brought from the West Indies; we have one of the cicada from the pupa, as well as others produced from wasps and bees in the perfect or winged state. Mr. Drury had a beetle in the perfect state, from every part of which small stalks and fibres have sprouted forth; they are very different from the tufts of hair that are observed on a few coleopterous insects, such as the Buprestis fascicularis, of the Cape of Good Hope, and are certainly a vegetable production.

ORTHOPTERA.

his plant, on which authors lay much stress, might be green, though a colour not so predominant in that tribe of vegetables as some others.*

The largest and most interesting of the Indian species of Mantis is found in the isle of Amboyna. Stoll contradicts the account of *Renard*,† who says these creatures are sometimes thirteen inches in length; but we have a specimen almost of that size.‡ It is related by Renard, and others, that the larger kinds of Mantes go in vast troops, cross hills, rivers, and other obstacles that oppose their march, when they are in quest of food. If they subsisted entirely on vegetables, a troop of these voracious creatures would desolate the land in their excursions; but they prefer insects, and clear the earth of myriads that infest it: if these become scarce from their ravages, they fight and devour one another. When they attack the plants, they do great mischief. It is said of some Locusts and Mantes that the plants they bite wither, and appear as if scorched with fire: we have not heard of this pestilential property in any of the larger species of Mantes.

Of the smaller kinds, the Mantis Oratoria is the most widely diffused, being found in Africa and Asia as well as in all the warmer parts of Europe. These creatures are esteemed sacred by the vulgar in many countries, from their devout or supplicating posture. The Africans worship them; and their trivial names in many European languages imply a superstitious respect for them.§

England produces no species of this tribe. The entomologists in this country must consequently rely on the accounts of those, who have observed them in other parts of the world. We shall select a few remarks from Roesel's extensive description of Mantis Oratoria and Gongyloides, because, if we may presume from the analogy they bear in form to Mantis Flabellicornis, the history of one will clearly elucidate that of the other.

Roesel says, some of the Mantes are local in Germany; they are found chiefly in the vintages at Moedting in *Moravia*, where they are called *Weinhandel*.‖ The males die in October, the females soon after.¶ The young brood are preserved in the egg state, in a kind of oblong bag, of a thick spongy substance; this bag is imbricated on the outside;

---

* These arguments of Donovan, although sufficiently ingenious, prove only the accidental possibility of the insect producing plants, and not the transformation which Piso believed to be the ordinary nature of the creature. (I. O. W.)

† Poissons des Molucques par M. Renard, Amsterl. 1754.

‡ Donovan here evidently alludes to some of the Phasmidæ.

§ Louva Dios by the Portuguese. Prie Dieu by the French.

‖ Probably a provincial term for a dealer in wine.

¶ Goetz, in his Beytrage, observes, that they live sometimes ten years.

it is fastened lengthwise to the branch of some plant.* As the eggs ripen they are protruded through the thick substance of the bag, and the larva, which are about half an inch in length, burst from them. Roesel, wishing to observe the gradual progress of these creatures, to the winged state, placed the bag containing the eggs in a large glass, which he closed, to prevent their escape. From the time they were first hatched they exhibited marks of a savage disposition. He put different sorts of plants into the glass, but they refused them, preying on one another: this determined him to supply them with other insects to eat: he put *ants* into the glass to them, but they then betrayed as much cowardice as they had barbarity before; for the instant the Mantes saw the ants they tried to escape in every direction. By this Roesel found the ants were the greatest persecutors of the Mantes. He next gave them some of the common musca (house flies), which they seized with eagerness in their fore claws, and tore in pieces: but, though these creatures seemed very fond of the flies, they continued to destroy one another through savage wantonness. Despairing at last, from their daily decrease, of rearing any to the winged state, he separated them into small parcels in different glasses; but here, as before, the strongest of each community destroyed the rest.

Another time, he received several pairs of Mantes in the winged state; profiting by his former observation, he put each pair [a male and female] into a separate glass, but they still shewed signs of an eternal enmity towards one another, which neither sex nor age could soften; for the instant they were in sight of each other, they threw up their heads, brandished their fore legs, and waited the attack: they did not remain long in this posture, for the boldest throwing open its wings, with the velocity of lightning, rushed at the other, and often tore it in pieces with the crockets and spines of the fore claws. Roesel compares the attack of these creatures to that of two hussars; for they dexterously guard and cut with the edge of the fore claws, as those soldiers do with their sabres, and sometimes at a stroke one cleaves the other through, or severs its head from the thorax. After this the conqueror devours his vanquished antagonist.†

We learn from Roesel also, the manner in which this creature takes its prey, in which respect we find it agrees with what is related of the extra European species. The patience of this Mantis is remarkable, and the posture to which superstition has attributed devotion, is no other than the means it uses to catch it. When it has fixed its eyes on an

---

* To that of the vine by Mantis Oratoria.

† The Chinese take advantage of these savage propensities, and keep these pugnaceous insects in little bamboo cages, training them to fight for prizes, as cocks are fought in this country. This custom is so common, that, according to Mr. Barrow, (Travels in China) " during the summer months, scarcely a boy is to be seen without his cage of these insects." (J. O. W.)

insect, it very rarely loses sight of it, though it may cost some hours to take. If it sees the insect a little beyond its reach, over its head, it slowly erects its long thorax, by means of the moveable membranes that connect it to the body at the base; then, resting on the four posterior legs, it gradually raises the anterior pair also; if this brings it near enough to the insect, it throws open the last joint, or crocket part, and snaps it between the spines, that are set in rows on the second joint. If it is unsuccessful it does not retract its arms, but holds them stretched out, and waits again till the insect is within its reach, when it springs up and seizes it. This is the uncommon posture before alluded to. Should the insect go far from the spot, it flies or crawls after it, slowly on the ground like a cat, and when the insect stops, erects itself as before. They have a small black pupil or sight which moves in all directions within the parts we usually term the eyes, so that it can see its prey in any direction without having occasion to disturb it by turning its head.

The most prevalent colour of this tribe of insects is fine green, but many of these fade or become brown after the insect dies: some are finely decorated with a variety of vivid hues; the most beautiful of these that we have seen are from the Moluccas.

TRUXALIS CHINENSIS.

Plate 10. fig. 1.

Section. Saltatoria, *Latreille*. (Gryllus, *Linn*.)
Family. Locustidæ. (Acridites, *Latreille*.)
Genus. Truxalis, *Fabricius*. (Gryllus, Acrida, *Linnæus*.)
Ch. Sp. Tr. viridis, capite thoraceque vittis quatuor, elytrorum lineâ centrali sanguineis, alis albido hyalinis. Expans. alar. 5¼ unc.
Tr. green, with four longitudinal stripes on the head and thorax, and a central line along the tegmina pink, wings stained pale buff hyaline. Expanse of the wings 5¼ inches.
Syn. Truxalis Chinensis, *Westw.*
Gryllus nasutus, *Donovan, 1st Edition*.

Donovan considered this insect as a variety of the Linnæan Gryllus nasutus, which he states to be found in Africa, Asia, and the south of Europe; adding, its varieties are numerous; and in size and colour depend on the climate they breed in. Sulzer represents it with red wings: in the Chinese specimens these are tinged with green. As several species are thus evidently confounded together, I have separated that here figured under the specific name of T. Chinensis.

2. Truxalis vittatus. 1. Truxalis Chinensis.

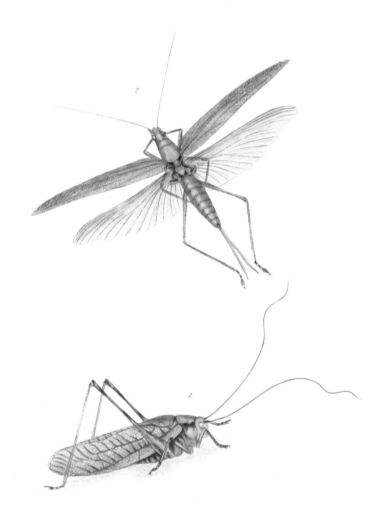

1. Gryllus perspicillatus. 2. Gryllus acuminatus.

TRUXALIS (MESOPS) VITTATUS.

Plate 10. fig. 2.

Sub-Gen. Mesops, *Serville*.

Ch. Sp. Tr. capite prominulo, testaceus, capite thorace femoribusque posticis vittâ laterali argenteâ. Long. Corp. $1\frac{1}{3}$ inch.

Tr. with the head slightly prominent; of a testaceous colour, with the head, thorax and posterior femora marked with a lateral silvery stripe. Length of the body $1\frac{1}{3}$ inch.

Syn. Truxalis vittatus, *Fabr. Ent. Syst.* 2. p. 27.

A single specimen of this insect, brought from China, and in the possession of Mr. Francillon, was employed as the original of this figure. From the form of the head it appears to belong to Serville's sub-genus, Mesops, but the wings extend beyond the body.

GRYLLUS (PHASGONURUS) PERSPICILLATUS.

Plate 11. fig. 1.

Family. Gryllidæ. (Gryllus Tettigonia, *Linnæus*.)
Genus. Gryllus, *Linnæus*. (Locusta, *Latreille*.)
Sub-Gen. Phasgonurus, (*Westw. Steph.* Locusta, *Serville*.)

Ch. Sp. G. capite pallido, antennis fuscis, thorace virescenti postice rotundato, elytris concavis viridibus nervosis; basi ocello dorsali fenestrato. (Long. Corp. elytr. claus. $2\frac{1}{4}$ unc.

G. with the head pale, antennæ brown, thorax green, rounded behind, elytra concave, green, nervose, with a dorsal fenestrated ocellus at the base. Length, with the wings closed, $2\frac{1}{4}$ inches.

Syn. Locusta perspicillata, *Fabricius Ent. Syst.* 2. p. 36.

Donovan states that Fabricius erroneously describes this insect as a native of America, and that it is not figured elsewhere. Fabricius refers to Dr. Hunter's Museum, now belonging to the University of Glasgow.

GRYLLUS (CONOCEPHALUS) ACUMINATUS?

Plate 11. fig. 2.

Sub-Gen. Conocephalus, *Leach, Serville.*
Ch. Sp. G. "thorace rotundato, vertice subulato, alis virescentibus." (*Linnæus loc. cit. subtus.*) (Expans. alar. fig. Donov. 4 unc.)
G. with the thorax rounded, forehead pointed, wings green. Expanse of the wings, according to Donovan's figure, 4 inches.
Syn. Gryllus acuminatus? *Linn. Syst. Nat.* 2. 696. *Fabr. Ent. Syst.* 2. p. 39. *Serville Revis. Orthopt.* p. 52. (Conocephalus ac.)

Donovan states that the insect here figured inhabits China and every other part of India. Linnæus, however, referring to Gronovius and Sloane's Jamaica, gives America as the locality of his G. acuminatus; whilst Fabricius says, "Habitat in omni India et in Europa Australi." In this confusion of locality, I have thought it best to mark the specific name of this figure with doubt.

LOCUSTA (RUTIDODERES) FLAVICORNIS.

Plate 12. fig. 1.

Family. Locustidæ, *Leach.* (Acridites, *Latreille.*)
Genus. Locusta. (Acrydium, *Latreille.*)
Sub-Gen. Rutidoderes, *Westw.* Acridium, *Serville.*
Ch. Sp. L. thorace subcarinato; viridis, elytris immaculatis, alis basi rufis, tibiis posticis sanguineis flavo-serratis. Expans. alar. 4 unc.
L. with the thorax somewhat keeled; green, with the elytra not spotted, wings at the base red, posterior tibiæ sanguineous with yellow teeth. Expanse of the wings 4 inches.
Syn. Gryllus flavicornis, *Fabricius Ent. Syst.* 2. p. 52.

1. Locusta flavicornis 2. Gryllotalpa Chinensis

Locusta morbillosa.

ORTHOPTERA.

GRYLLOTALPA CHINENSIS.

Plate 12. fig. 2.

FAMILY. ACHETIDÆ. (Gryllus Acheta, *Linn.*)
GENUS. GRYLLOTALPA, *Latreille*.
CH. SP. G. luteo fulvescens, tibiis anticis 4-dentatis. Long. Corp. 1 unc.
G. of a fulvous-clay colour, anterior tibiæ with four teeth. Length 1 inch.
SYN. G. Chinensis, *Westw.*
Gryllus Gryllotalpa, *Donovan*, 1st edition.

The genus of Mole-crickets, Gryllotalpa, comprises several distinct, although very closely allied species, the inhabitants of various countries, differing especially in the number of teeth of the anterior tibiæ and the neuration of the tegmina. Donovan regarded them as varieties only, observing that the species here figured "differs in no respect from the European species except in size and colour," the English mole-cricket "being twice as large and of more of a mouse colour."

LOCUSTA (PHYMATEA) MORBILLOSA.

Plate 13.

FAMILY. LOCUSTIDÆ.
GENUS. LOCUSTA.
SUB-GEN. PHYMATEA, *Westw.* Phymateus, *Thunberg, Serville.*
CH. SP. L. thorace quadrato, rubro, verrucoso; elytris fuscis, flavido punctatis; alis rufis, nigro punctatis. Expans. alar. 4⅓ unc.
L. with the thorax squarish, red, warty; elytra bluish-brown, with pale yellow spots; wings red, with black spots. Expanse of the wings 4⅓ inches.
SYN. Gryllus morbillosus, *Linn. Syst. Nat.* 2. 700. (e Cap. Bon Spei). *Fabricius Ent. Syst.* 2. *p.* 50. *Serville Revis. Orthopt. p.* 86. (Phymateus m.)

Donovan states that "the Gryllus morbillosus appears in the early edition of the *Systema Naturæ* and the works of Roesel as an Indian species, and that Mr. Drury assured him he had received it several times from China. Another sort is also found at the Cape of Good Hope, which is rather larger and deeper in colour than the Chinese variety."

If this be correct, it will, I apprehend, be necessary to consider these two *sorts* as distinct species, to retain the specific name morbillosus for the African species, and to give a new name to the Chinese species.

When this insect is at rest, the wings are folded and much of its beauty is concealed; but when these are expanded, its appearance is altogether magnificent. It has nothing of the shining and metallic splendour of the Coleoptera, for its colours are translucent, and assume their richest hues when they pass before the light. The elytra are purple, variegated with yellow; the wings of a glowing crimson, spotted with black; the abdomen is surrounded with alternate zones of black and yellow, and the legs are throughout of an elegant scarlet, inferior only in brightness to the coral red of the head and thorax. Upon the whole, this species is embellished with such a profusion of various and beautiful colours, that it may be considered as a most splendid example of the Linnæan Hemipterous order of insects. It is represented on the *Iris Chinensis* in a flying position.

This is not supposed to be a numerous species in China; on the contrary, it is probably uncommon. Several others of the locust are abundant in that country, and in seasons favourable to their increase do incredible mischief.* Both the Locusta tartarica and Locusta migratoria inhabit Tartary on the northern confines of China, from whence, at certain periods, they descend like an impetuous torrent over the neighbouring countries in quest of food, strip the earth of verdure, and scarcely leave the vestige of vegetation behind them. The Locusta migratoria, whose myriads are said to darken the face of heaven in their flights, sometimes direct their course westward, cross rivers, sea, and an immense extent of country, till they reach Europe; and though many are lost in these bold migrations, the survivors are in sufficient numbers to commit vast depredations. This species has been known to visit England,† but not in any abundance. In Little Tartary and the European provinces of Turkey, in Italy, and in Germany,‡ they do great mischief in these migrations. The Locusta flavicornis and

---

* " Famines sometimes happen in this part of the province; in some seasons inundations produced by torrents from the mountains, and as often the depredations of locusts, are causes of this disaster." (*Sir J. Staunton, Chap. on Tien-sing.*)

† The last appearance of this species in England was in 1748. Donovan had specimens of it from Smyrna, Germany, and China, and deemed it too common and general an inhabitant to merit a figure as a Chinese insect.

‡ Roesel speaks of this locust infesting the provinces of Wallachia, Moldavia, and Transylvania, in such immense numbers in the years 1747, 1748, and 1749, that an *Imperial and Royal Hungarian edict* was issued, with printed instructions for the best means of exterminating them. (*Der Heuschrecken-und Grillensammlung, &c. &c. vol. II. page* 193.)

Fulgora Candelaria.

nasuta are two other abundant species in China, and no doubt there are many other common kinds in that country we are at present unacquainted with. The locust is only detrimental when in immense numbers; for in China, as in other eastern countries, they are considered as an article of food, and regularly exposed for sale in the public markets.*

Order. **HEMIPTERA.** *Linnæus, Latreille.*

FULGORA CANDELARIA.

Plate 14.

SUB-ORDER.	HOMOPTERA, *Latreille.*
FAMILY.	FULGORIDÆ.
GENUS.	FULGORA, *Linnæus.*
CH. SP.	F. fronte rostratâ, adscendente; elytris viridibus luteo-maculatis; alis flavis apice nigris. Expans. alar. 3 unc.
	F. with the forehead produced into an ascending rostrum; elytra green, spotted with luteous; wings yellow buff, with the tips black. Expanse of the wings 3 inches.
SYN.	Fulgora candelaria, *Linn. Syst. Nat.* 2. 703. *Roesel Ins.* 2. 189. *t.* 20. *Sulzer Ins. t.* 10. *f.* 62. *Fab. Ent. Syst.* 4. *p.* 2. *Syst. Rh. p.* 2. *De Geer Ins.* 3. 197. 2. *Act. Holm.* 1746. *t.* 1. *f.* 5. 6.

The phenomena resulting from the properties and effects of light having engaged the attention of the earliest philosophers, we must conclude that phosphorical appearances, and those especially of animated bodies, could not fail to attract their particular notice. Indeed it is evident from the writings of the accurate observers of nature in remote ages, that they were acquainted with certain insects that have the property of shining in the night. These were known only by general terms, expressive of that property; yet it is probable that some of the Linnæan Lampyrides, which are abundant in the south of Europe, as well as in Asia and some parts of Africa, were the first of the illuminated

---

* Sir G. Staunton likewise speaks of "a large species of Gryllus" that is kept in cages for amusement in China, and was exposed for sale with other insects in the shops of Hai-ten. Neither the species of this, or the locusts noticed in the preceding note, are mentioned. (This remark evidently applies to the Mantidæ mentioned in the note to Empusa flabellicornis.)

insects known to them.* Some of the males, which are furnished with wings and are illuminated like the females, were striking objects of natural history, and could scarcely have escaped their notice. The Greeks included all shining insects under the name lampyris; and the Latins called them cicindela, noctiluca, luciola, lucernata, &c. Whether any of the *Fulgoræ* were known to the ancients is uncertain; probably they were not, the most remarkable species being peculiar to the warmest parts of America. Asia, once the seat of learning, does indeed produce a few species; but we have no account of these in ancient natural history.

The Fulgoræ seem to have been entirely unknown in Europe till the latter end of the seventeenth century, when two writers published descriptions and figures of *Fulgora Laternaria;* Madame Merian, of Holland, in her splendid work on the Metamorphoses of the Insects of Surinam, and Dr. Grew, of London, in his Rarities of Gresham College.

Reaumur† is the next author who described the *Fulgora Laternaria*, and after him Roesel, in his " Amusing History (or Recreation) of Insects."‡ This brings us to the period in which *Fulgora Candelaria*, our Chinese species, was first known in Europe; a circumstance of much importance to naturalists at that time, because the first-mentioned species was a solitary example of its singular genus. The transactions of the Stockholm academy includes the earliest figure and description of this extraordinary insect.

Roesel has given three figures and a description of it, and from his account we learn that it was known in England before he was acquainted with it. On its peculiar qualities he had been unable to derive any information concerning it, but his description is notwithstanding extremely prolix. We have selected the most interesting passage, because it clearly marks the progressive advancement of the knowledge of natural history in Europe so late as the middle of the present century.

" According to my promise," says Roesel,§ " I now produce the second sort of Lantern-carrier, which I never saw before, and of which I have never read in any work on insects. The scarcer, however, it may be, the more I am indebted to Mr. Beurer, apothecary of this place,|| &c. for the permission he has granted me to draw and enrich my collection with it. Mr. Collinson has sent it to him from London, under the name

---

* The lampyris of *Pliny* is expressly the insect with a shining tail.
† Memoires pour servir a l'Histoire des Insectes. 1734.
‡ Insecten Belustigung.
§ Versprochener massen liefere ich nunmehr die zweite Sorte des Lanternen-Tragers, &c. Vol. i. pl. 30. Locust, page 189.
|| Nurenberg.

Lanternaria Chinensis, for which reason I have called it the Asiatic or Chinese Lantern-carrier." Roesel being a respectable entomological writer of his time, we must infer that *Fulgora Candelaria* was extremely scarce in Europe when his plate and descriptions were published. The commercial concerns of Europeans with the Chinese having greatly increased since that period, has facilitated many inquiries concerning the natural productions of China; and amongst a variety of other insects that are now usually brought from that country, specimens of Fulgora Candelaria are extremely common. In China, few insects are found in greater abundance.

Having noticed the early history of this insect, we come to consider the peculiar properties of its singular genus; upon which the following observations were made by our author: "Among these we find the most astonishing that insects can possess, that of emanating light; not merely a momentary shining appearance, as is produced by many viscous substances, but a clear and constant resemblance to the element fire, and capable of diffusing light to surrounding objects, though totally destitute of every principle that can do mischief. To the unphilosophical mind it appears at first impossible, and it cannot fail to astonish the best informed; indeed, some readers might be inclined to doubt the veracity of travellers in foreign countries who have seen a vegetable* or an animal produce light, if our own country could not supply us with abundant analogous proofs of such phenomena. The presence of this animated phosphorus, if we may so express it, is observed on several insects that are natives of England; it is needless to enumerate them, because the most striking example must be recollected by every rural inhabitant or admirer of poetical simplicity:

> '———————— On every hedge
> The glow-worm lights his gem, and through the dark
> A moving radiance twinkles.'   THOMSON.

"The account which Madame Merian gave of the effect of the light produced by the *Fulgora Laternaria*† was greatly discredited, though Dr. Grew had related some sur-

---

* An instance of this occurs in the south of Europe. An account in the Philosophical Transactions relates of the Dictamnus Albus (Fraxinella), that " in the still evenings of dry seasons it emits an inflammable air, or gas, and flashes at the approach of a candle. There are certain instances of human creatures who have taken fire spontaneously, and been totally consumed."

† The account which Madame Merian has given of the light of the *Fulgora Laternaria* is so surprising, that it will certainly prove acceptable to many readers. It is indeed a digression from the account of *Fulgora Candelaria*, but will tend to prove, that insects of this genus emit a more vivid light than any of the illuminated kinds hitherto known.

" Once," says Madame Merian, " when the Indians brought me a number of these Lantern-carriers, I put

HEMIPTERA.

prising particulars of a specimen of it from Peru.* Her account has, however, been generally believed since the missionaries† in countries which produce those insects have confirmed her account. It is admitted that the *Chinese Fulgora* has an illuminated appearance in the night. 'The foreheads of many Fulgoræ (especially those found in China) emit a lively shining light in the night-time, which, according to some authors, is sufficient to read by.'—*Yeats.* ‡

"The light of the Fulgoræ is generally imagined to issue from the trunk, or elongated projection of the forehead; but Roesel offers a conjecture on the light of the Fulgora Laternaria, which, on further investigation, may enable naturalists to determine whether the light is entirely produced by an innate property of the trunk, or receives additional splendour from some external cause. He notices a white farinaceous substance on several parts of the wings and body as well as the trunk, which, he observes, looks like the decayed wood which shines at night. We mention this conjecture of Roesel, though the same occurred to us before we perused his observations. We have invariably found a similar white powder on other insects of this genus, but usually upon the trunk only. The remarks of Roesel were necessarily very limited, two species of the Fulgoræ only being then known. We possess twelve distinct species, and have made dissections and observations on several others; from all which we are inclined to imagine

---

them into a wooden box, without being aware of their shining at night; but one night, being awakened by an unusual noise, and much frightened, I jumped out of bed and ordered a light, not knowing whence this noise proceeded. We soon perceived that it originated in the box; we opened with some inquietude, but were still more alarmed after opening it, and letting it fall on the ground, for a *flame* appeared to issue from it, which seemed to receive additional lustre as often as another insect flew out of it. When we observed this some time we recovered from our terror, and admired the splendour of these little animals." *Dissert. de generatione et metamorphibus Insectorum Surinamensis.*

* Cucujus Peruvianus. "That which, beside the figure of the head, is most wonderful in this insect, is the shining property of the same part, whereby it looks in the night like a lantern, so that two or three of these fastened to a stick, or otherwise conveniently disposed of, will give sufficient light to those who travel or walk in the night." *Grew. Museum Regalis Societatis, p.* 158.

† Le ver-luisant. Ceux que nous voyons à la campagne dans les nuits d'été ne jettent qu'une foible lueur: mais ils y en a dans les Indes modernes qui répandent un éclat très-vif. Ce sont, pour ainsi dire, des phosphores animez. "Les Indiens," dit le savant auteur de la Théologie des Insectes, "ne se servoient autrefois dans leurs maisons, et dehors d'aucune autre lumière. Lorsqu'ils marchent de nuit, ils en attachent deux aux gros doigts du pied, et en portent un à la main. Ces insectes répandent une si grande clarté, que par leur moyen on peut lire, écrire, et faire dans une chambre toutes les autres choses nécessaires." *Lesser. Liv.* 2. c. 3. rem. 8. "Le trait rapporté par le P. DU TERTRE dans son *Histoire des Antilles*, auroit bien dû être cité, il dit avoir lû son bréviare à la clarté d'un de ces vers-luisans."

‡ Yeats. Institutions of Entomology.

Cicada Atrata.

HEMIPTERA.

that the white powder has a phosphoric appearance in the living insect, and increases the light when the end of the trunk is illuminated.

"One of the Fulgoræ of considerable magnitude, from the interior of India, enabled us to make many observations. The trunk is of the same form as that of the *Fulgora Candelaria*; the colour is a dark but beautiful purple; the apex scarlet, of a perfectly pellucid appearance, and still retains a reddish glare; the spots of white sprinkled on the purple colour of the trunk exhibit also a slight appearance of phosphoric matter. On the trunk of the Fulgora Candelaria these white spots are very conspicuous.

"Though the generic name Fulgora seems to imply some effulgent property in the insects that compose the genus, it is uncertain whether all possess that property. They are indeed furnished with a trunk, but it is smaller in proportion in several species than in F. Laternaria, Candelaria, Flammea, Phosphorea, and some others. It has not been determined whether any of the European Fulgoræ shine in the night-time." *

CICADA ATRATA

Plate 15.

FAMILY. CICADIDÆ.
GENUS. CICADA, *Linnæus*. Tettigonia, *Fabricius*.
CH. SP. C. atra, alis albis basi nigris, venis testaceis. Expans. alar. 4¾ unc.
C. black, with iridescent white wings black at the base, and with yellowish brown nerves. Expanse of the wing 4¾ inches.
SYN. Cicada atrata, *Fabr. Ent. Syst.* 4. p. 24. no. 28. *Syst. Rhyng.* p. 42.

Though the observations of Sir G. Staunton on the natural productions of China were necessarily very general, the study of insects appears to have engaged his particular attention; and on that account we must lament that untoward events precluded him from observing more minutely the peculiarities of some kinds, and the economical

---

* (Notwithstanding the observations of Donovan and the various authorities cited by him, it is certainly a matter of doubt, at the present day, whether the Fulgoræ in reality possess any luminous property. No recent author or traveller has noticed its existence in these insects, although it is related in every work of travels as exhibited by the Elateridæ and Lampyridæ. Moreover, the farinaceous matter noticed by Rösel exists in many other insects known to be not in the least degree luminous, and of which the woolly or waxen appendages in Lystra lanata, the Dorthesiæ, Cocci, &c. is but a modification or more extensive developement. M. Wesmael, of Brussels, has, however, just reasserted the luminous properties of the South American Fulgora on the authority of a friend who had witnessed it alive. *Ann. Soc. Ent. France*, 1837, *App.* J. O. W.)

purposes of others. We therefore peruse the following account of an unknown species of Cicada with particular regret, because it withholds information interesting to the naturalist, and from its air of novelty is likely to promote an erroneous opinion concerning that singular tribe of insects.

" The low and sometimes marshy country through which the river* passes, is favourable to the production of insects; and many of them were troublesome, some principally by their sting, and others by their constant stunning noise. The music emitted by *a species of Cicada* was not of the vocal kind; but produced by the motion of two flaps or lamellæ which cover the abdomen or belly of the insect. It is the signal of invitation from the male of *that species* to allure the female, which latter is quite unprovided with these organs of courtship."†

Again, when describing a town higher up the river, that author says, " The shops of Hai-tien, in addition to necessaries, abounded in toys and trifles, calculated to amuse the rich and idle of both sexes, even to cages containing insects, such as the *noisy Cicada*, and a large species of the Gryllus."‡

The reader may imagine from the first account, that the music of every other species of Cicada is of the vocal kind, or that it is peculiar to this Chinese insect to be furnished with lamellæ that cause a sound. The latter account confirms such conjecture, by alluding in a specific manner to the *noisy Cicada*, as to an insect described in a former part of the work. We must remark, that not only the males of the species mentioned by that author are furnished with those lamellæ, but the whole of that section of the Linnæan Cicadæ which Fabricius has called Tettigonia. The males of the species included in the other sections of that genus are certainly furnished with them also; though some of them are too minute to be observed without a glass. These lamellæ vary in size in different species; but the accounts we have of them from travellers in foreign countries, and naturalists both ancient and modern, prove that they all emit a certain sound to allure the female. As we are unable to ascertain the Chinese species Sir George mentions, neither figure nor description accompanying his account of it, we must therefore speak generally of the whole genus, and then confine our remarks to those species we are acquainted with from China. Among these are *C. splendidula, sanguinea,* and *atrata*. The latter, we believe, is the largest species of the Chinese Cicadæ known in Europe.

---

* Pei-ho.
† Chap. 3. Vol. II. octavo.
‡ Chap. 4. Vol. II.

HEMIPTERA.

Some species of this tribe were known to the ancients. With them it was the emblem of happiness and eternal youth;* and if we examine the legends of pagan mythology, we find they were deemed a race of creatures beloved by gods and men. The Athenians wore golden Cicadæ in their hair, to denote their national antiquity; or that like those creatures they were the *first born* of the earth; and the poets feigned that they partook of the perfection of their deities.† Anacreon depictures in glowing colours the uninterrupted felicity of this creature: his ode to the Cicada is appropriate to our inquiry.‡

In the infant state of music, men seem to have preferred the natural sounds of some animals to those of their uncouth instruments. We cannot otherwise account for the extravagant praise bestowed on the noise of this little creature. It is true, authors agree that the sounds of some kinds are exceeding loud and harmonious, and in the early ages of the world these might have a powerful influence on the human mind. It is related that the *ancient Locri*, a people of Greece, were so charmed with the song of the Cicada, that they erected a statue to its honour.§

The ancients had attentively observed the manners of its life, though they indulged

---

* Probably because it was supposed to live only a short time. The renewal of youth is illustrated by the story of the Tithonus transformed by Aurora into a Cicada.

† These pagan deities were without flesh or blood, and composed of aerial and watery humours. Such they imagined the moisture of the Cicada, and perhaps for that reason first assigned it a place among their demi-gods.

‡ Happy creature! what below
   Can more happy live than thou?
   Seated on thy leafy throne,
   (Summer weaves the verdant crown,)
   Sipping o'er the pearly lawn
   The fragrant nectar of the dawn;
   Little tales thou lov'st to sing,
   Tales of mirth—an insect king:
   Thine the treasures of the field,
   All thy own the seasons yield;
   Nature paints for thee the year,
   Songster to the shepherds dear:
   Innocent, of placid fame,
   What of man can boast the same?

Thine the lavished voice of praise,
Harbinger of fruitful days;
Darling of the tuneful nine,
Phœbus is thy sire divine;
Phœbus to thy notes has giv'n
Music from the spheres of heav'n:
Happy most, as first of earth,
All thy hours are peace and mirth;
Cares nor pains to thee belong,
Thou alone art ever young;
Thine the pure *immortal* vein,
Blood nor flesh thy life sustain;
Rich in spirits—health thy feast,
Thou'rt a demi-god at least.
    Green's Trans. Ode 43.

§ Some say, that once a certain player of *Locri*, contesting in the art of music with another, would have lost the victory, by the breaking of two strings of his instrument, but a *Cicada* flew to his aid, and resting on the broken instrument, sung so well, that the Locrian was declared victor. The Locrians erected a statue to the Cicada as a testimony of their gratitude. It represented the player with the insect on his instrument.

F

in many poetical fictions concerning it; and particularly when they affirmed that it subsisted on dew. They have told us that it lives among trees,* which circumstance discountenances the opinion of those moderns who imagine the grashoppers† were the Cicadæ of the ancients.

Neither were they ignorant that the males only were furnished with those instruments which externally appear to produce its sound, or the purpose for which that sound was emitted;‡ though it was reserved for more accurate naturalists to discover the complex organs by which it was caused and modulated. Aldrovandus, near two centuries ago, described the lamellæ, which he compares to the fruit of some herbs, called by modern botanists *Thlaspi*.§

Among later naturalists who have noticed the Cicadæ of foreign countries, are Merian,‖ Margravius,¶ &c. Merian says, its tune resembles the sound of a lyre which is heard at a distance; and that the Dutch in the plantations of Surinam (where they are very plentiful) call it the Lyre-player.** Margravius, in his natural history of Brazil, compares it to the sound of a vibrating wire: he says the tune begins with *Gir, guir*, and continues with *Sis, sis, sis*. One species is called Kakkerlak†† in the Indies, perhaps because the sound emitted by it may be likened to the pronunciation of that word. Mr.

---

* Dr. Martyn supposed this refers to the smaller branches in hedges, rather than to the lofty trees in forests: we cannot entirely coincide with that opinion.

† *Grashopper. Cicada.* They live almost every where in hot countries. *Lovel. Hist. Animal.* containing the summe of all authors ancient and modern, *p.* 274, &c. &c.

Cicada, a Sauterelle,ᵃ or, according to others, a *balm cricket*.—Non est quod vulgò, a *grashopper*, vocamus; sed insectum longè diversum, corpore et rotundiore et breviore, qui arbusculis insidet et sonum quadruplo majorum edit. a *grashopper*, rectè locustum reddideris, *Morl* ex *Ray*. *Ainsworth*.

‡ Xenarchus, an old Grecian play-writer, used to say jocosely, that "the Cicadæ were very happy because they had silent wives." Aristotle also knew the sexual difference of them; he mentions them as a delicious food: he preferred the males when young, but more so the females before she laid her eggs.

§ Thalaspi parvum Hieraciifolium, sive Lunariam luteam Monspel. et Leucoium luteum marinum. *Lobel. Stirpium Adversaria nova, p.* 74.—*Aldrov.*

‖ Merian. Insecta Surinamensia.

¶ Georgi Margravii rer. nat. Brasiliæ. *Lib.* 7. *p.* 257.

** De Lierman.

†† *Scopoli,Carn.* Yeats describes the Kakkerlak of the American islands as a species of *Blatta*, cock-roaches. Are there not two insects of that name?—one of them is, we believe, a *Blatta*. Indeed, Latreille has made use of the name Kakkerlak for a genus of Blattidæ.

ᵃ *Sauterelle*, sorte d'insecte. A locust or grashopper. *Boyer.*—*Cigale*, a flying insect. The Cicada of the ancients, unknown in England. *Boyer.*

Abbot, an accurate observer and collector of natural history in *North America*, has discovered four new species of Cicada, one of them nearly equal in size to our Cicada Atrata. This, he says, was found in great abundance one season in some swampy grounds near the Susquehanna river, and was remarkable on account of its loud noise, which at a little distance resembled the ringing of *horse-bells*.*

Some naturalists have supposed that the sound of the Cicada is caused by the flapping of the lamellæ against the abdomen; and others, that it is only a noise occasioned by the rustling of the segments of the body in the contractile motion of that part. Beckman† imagines it is caused by beating the body and legs against the wings: he has endeavoured to explain the meaning of ancient authors, and deduce its etymology from that circumstance.‡

Reaumur and Roesel have dissected several of the Cicadæ, and discovered that the lamellæ cannot have that free motion necessary to cause such a sound; but that it is produced by some internal organs of the insect, and only issues through the opening concealed under the lamellæ as through the mouth of a musical instrument.§

---

* Communicated by Mr. Abbot, in North America, to Mr. Francillon, in London.

† *Roes. Insecten Bellustigung.*—CHRISTIANI BECKMANNI, Bornensis, manuductionem ad latinam linguam: nec non de originibus latinæ linguæ, &c.

‡ It is the common opinion that the word *Cicada* has its origin from *quod cito cadat*, which, after a general interpretation, implies that the Cicadæ *soon vanish*, or are *short-lived*. Beckman maintains that this opinion is absurd, and proves that its name is derived from singing, because φ ἄδειν signifies a sound produced by the motion of a little skin; and that *ciccum* or *cicum* is a thin little skin of a pomegranate that parts the kernels.—Beckman not knowing the insect, or not imagining that the *little skin* was an appendage to the abdomen, concluded it must mean the transparent wings, and consequently that the sound was produced by beating them against the body: but this interpretation, if applied to the lamellæ instead of the wings, will directly prove the origin of its name, and knowledge of the ancients.

§ For the satisfaction of the curious reader, we detail the most interesting particulars concerning the organization of these parts from *Reaumur's Histoire des Insectes*, and *Roesel's Verschiedene auslaendische sorten von Cicaden, &c.*

The music of the Cicada is not caused by the motion of the *lamellæ* as some have supposed. Reaumur observes, that although the *lamellæ* have a kind of moveable hinge, they have also a stiff and pointed tooth, or spine, that prevents them from being lifted far back; and if strained are very liable to be broken.

From the anatomical description of Roesel, we find that, within the two hollows that are seen when the lamellæ are lifted up, two very smooth skins are visible; these are highly polished, of nearly a semicircular shape, and reflect prismatic colours: there is between these a hard brown projection, or corner which unites with another piece above them in a longitudinal direction, to the under part of the breast. This longitudinal piece divides a triangular red space or field into two parts, one on the right side, and the other on the left. Above these, in a transverse direction, are seen two small yellow skins; the lamellæ in their natural position conceal these organs because they fold exactly over them.

Reaumur, in the exterior appearance of these parts, could discover nothing that could lead to determine the

HEMIPTERA.

The suppositions of these authors seem well founded; we have examined many species that were unknown to them, and find the spine before mentioned so placed in many insects as to prevent the motion of the lamellæ. We have a specimen from America, which, in addition to the usual organs of sound, have two large hollow protuberances or drums; one on each side of the abdomen; and must, we imagine, produce a louder sound than any yet discovered; a species very similar to this is also brought from New Holland.

The proboscis of these insects is a hard or horny tube, in which a very acute slender sucking-pipe is concealed. The horny tube is not unlike a gimlet in form, and is used by those creatures to bore through the bark of trees, to extract the juices on which it feeds. Linnæus has named the species of one division, in his System, *Mannifera*, because they had been observed to fly among ash trees, bore many holes in them, and when the manna had oozed out return and carry it off.

With this proboscis they bore holes in the small twigs of the extreme branches of trees and deposit their eggs in them, sometimes to the amount of six or seven hundred. As each cell contains no more than from twelve to twenty eggs, it does great damage to the trees they frequent. Stoll says, " the common one,* which is found at Surinam in the coffee plantations, greatly injures those trees; the females depositing their eggs in

---

organs of the sound; and he was not satisfied that the slight motion of the lamellæ on these parts could produce the loud singing noise of the Cicada. He opened a few cicadæ on the back part of the body, so that the inner structure of the under side was displayed, and especially the parts connected to the curious organs he had discovered under the lamellæ. At last he discovered two large muscles, which at their point of union formed a space almost square, and were connected with the red triangular fields he had observed on the under side: as he concluded these formed a material part of the organs he wished to discover, he examined them attentively, and found that, by moving them backwards and forwards, he could make a cicada sing that had been dead many months. Although the sound was not strong, it tended to prove that he had discovered the instrument that produced it.—In another part he says, it is evident the sound is caused by the little skins connected to the muscles, because when they were rubbed with a bit of paper they emitted that kind of sound.

Roesel has discovered two little pieces of horny substance that are connected by a sort of fibre within the skins, in the body, and he supposes when this is in motion, it strikes against the before-mentioned thin skins, and produces a sound, by the same means as a hollow body, or drum, when struck with a stick: and also that this noise may be varied or modulated by a slight motion of the lamellæ, but cannot be produced without the assistance of the internal nerves and muscles connected with the organs first described.

Authors agree that the Cicadæ of hot countries emit the loudest sound. It appears from the papers of Mr. Smeethman (who resided a considerable time in Africa) published by Mr. Drury, that the sound of some kinds peculiar to that part of the world is so loud as to be heard at half a mile distance: and that the singing of one within doors silences a whole company.—The same attentive observer says, the open parts of the country are never without their music, some singing in the evening, and others only in the day.

* La Cigale Vieilleuse. Cicada Tibicen.

the young shoots, and in holes they bore with their sheath. They live on the juices of the trees."

M. Merian gives a figure and account of the metamorphosis of a cicada found in Surinam. She has mistaken the winged insect to be only the pupa of the *Fulgora Laternaria*, which is too absurd to deserve contradiction; in other respects her account is interesting, and particularly that part which relates to the pupa state, or chafer, as it is termed. "The pomegranate tree," says Merian, "so well known in all other countries, grows also in the fields of Surinam. On them I have found a species of chafer, which is naturally very lazy, and consequently very easy to be caught. It carries underneath the head a long trunk, with which it easily penetrates the flowers, in order to extract the honey from them. On the 20th of May, when they were laying quite quiet, the skin of the back burst open, and green flies, with transparent wings, issued from them. These flies are found in abundance in Surinam, and have such a rapid flight, that it took me many hours to catch one."

The pupa Donovan received from China with Cicada atrata very much resembles that figured by Merian. It has the long sucking trunk or proboscis; but the most formidable of its weapons seem to be the fore feet, which are thick, strong, and armed with spines or teeth; with these it may do more injury to the plants, by tearing off the tender shoots, than by wounding the trunk to extract the moisture.

The upper and under side of a male of Cicada atrata are represented, not only to illustrate our preceding remarks, but because Donovan believed no figure had been given of it by any author, unless *De Zweite Chineesche cicade* of Stoll (Pl. 20. fig. 118.) is intended for this insect.

The general appearance of both sexes of Cicada atrata is very similar, except that the female is furnished with a sheath, and the male with lamellæ. The sheath of the female is partly concealed within a valve at the extremity of the abdomen, and is only protruded when the creature lays her eggs. In the figure of the under surface of a male insect, exhibited in the annexed plate, the lamellæ are distinguished by two stars: the single star denotes the situation of the spine, mentioned by Roesel and Reaumur.

The Camphor-tree, *Laurus Camphora*, is represented in the plate. The tree which produces the useful drug *camphor* is very abundant in Japan and China. Sir G. Staunton says, it is the only species of the laurel genus growing in China, where it is a large and valuable timber tree, and is never cut up for the sake of the drug; but that substance is obtained by decocting the small branches, twigs, and leaves, and subliming the camphor in luted earthen vessels. A purer sort is brought from the island of Borneo and Japan, which is supposed to be a natural exudation from the tree when the bark is wounded. Sir G. Staunton says, the Camphor-tree is felled in those countries for the sole purpose of finding the drug in substance among the splinters.

HEMIPTERA.

CICADA SANGUINEA.

Plate 16. fig. 1.

CH. SP. C. nigra, facie, thoracis maculis duabus, abdomineque sanguineis. Exp. alar. $2\frac{1}{6}$ unc.
C. black, with the face, two spots on the thorax, and the abdomen blood red. Expanse of the wings $2\frac{1}{6}$ inches.

SYN. Cicada Sanguinea, *De Geer Ins.* 3. 221. *tab.* 34. *fig.* 17. *Donovan, 1st edit.*
Tettigonia Sanguinolenta, *Gozen Entomol. Beitr.* II. *p.* 150. *Fabricius Ent. Syst.* 4. *p.* 25. *Germar in Silberm. Rev. Ent.*

The specific name sanguinea of De Geer having the priority, is here retained in preference to that of Fabricius (Ent. Syst.), which is identical with that of another species of the Linnæan Cicadæ, as well as with that which, in his Systema Rhyngotorum, he has proposed in lieu of C. hæmatodes of his earlier works, which is however distinct from the Linnæan C. hæmatodes, and for which it will be necessary to employ another specific name.

CICADA AMBIGUA.

Plate 16. fig. 2.

CH. SP. C. olivacea, elytris hyalinis, marginibus anticis testaceis. Expans. alar. $3\frac{1}{4}$ unc.
C. olive coloured, with the elytra clear, the anterior margin testaceous. Expanse of the wings $3\frac{1}{4}$ inches.

SYN. Cicada ambigua, *Donov. 1st edit.*

Received by Mr. Drury from China.

LYSTRA LANATA.

Plate 16. fig. 3.

FAMILY. FULGORIDÆ.
GENUS. LYSTRA, *Fabricius.*
CH. SP. L. elytris apice nigris, punctis cœruleis, fronte lateribusque rubris, ano lanato. Expans. alar. $1\frac{3}{4}$ unc.
L. with the elytra black at the tips, spotted with blue, with the front and sides of the head red, abdomen woolly at the extremity. Expanse of the wings $1\frac{3}{4}$ inch.

SYN. Cicada lanata, *Linn Syst. Nat.* 2. 711. *Fabr. Ent. Syst.* 4. 30. *Syst. Rhyng.* *p.* 56. *Sulzer Ins. t.* 9. *f.* 11. *Stoll Cicad. t.* 10. *f.* 49.

1. Cicada sanguinea.
2. Cicada ambigua.
3. Lystra lineata.
4. Cicada splendidula.
5. Cercopis abdominalis.
6. Jassus frontalis.

HEMIPTERA.

One of the most beautiful species of the Indian Cicadæ. The wing cases are black, elegantly reticulated, and spotted with bright blue. At the extremity of the abdomen it has a tuft of long and very delicate hairs, intermixed with others that are rather convoluted and of a coarser texture. The whole of this insect, but particularly between the abdomen and wings, is sometimes profusely covered with a fine powder of a snowy whiteness, similar to that observed on the Flata limbata in the imperfect state; hence we may conclude it is also one of those insects which furnish the white wax* so highly esteemed in China.

CICADA SPLENDIDULA.

Plate 16. fig. 4.

Ch. Sp.   C. elytris fusco-aureis, femoribus anticis incrassato-dentatis rufis. Long. Corp. alis clausis. $\frac{3}{4}$ unc.

C. with golden brown elytra, the anterior femora incrassated, toothed and red, thorax and scutellum varied with yellow and black. Length, with the wings closed, $\frac{3}{4}$ inch.

Syn.   Cicada splendidula, *Fabricius Ent. Syst.* 4. p. 25. *Syst. Rhyng.* p. 42.

Figured from the unique specimen in the collection of Mr. Drury, described by Fabricius.

CERCOPIS ABDOMINALIS.

Plate 16. fig. 5.

Family.   Cercopidæ.

Genus.   Cercopis, *Fabricius*.

Ch. Sp.   C. atra nitida, thorace immaculato, elytris basi fasciaque media flavescentibus; abdomine sanguineo. Long. Corp. alis clausis $\frac{1}{3}$ unc.

C. black shining, thorax without spots, elytra with the base and a central fascia yellowish, abdomen sanguineous. Length, with the wings shut, $\frac{1}{3}$ inch.

Syn.   Cicada abdominalis, *Donovan, 1st Edit.*
Cercopis Heros? *Fabr. Syst. Rhyng.* p. 89.

---

* Vide Sir G. Staunton's *Hist. Emb. China.*

TETTIGONIA FRONTALIS.

Plate 16. fig. 6.

GENUS.   TETTIGONIA, *Latreille, Germar.*
CH. SP. T. pallida, occipite thoraceque punctis quinque nigris, fronte puncto nigro inter oculos, elytris sanguineis. Epans. alar. fere 1 unc.

T. pale, with five black spots on the head and thorax, and one in front between the eyes, elytra red. Expanse of the wings nearly 1 inch.

SYN.   Cicada frontalis, *Donovan*, 1st Edit.
Cicada cæruleipennis? *Fab. Syst. Rh. p.* 73.

FLATA NIGRICORNIS.

Plate 17.

GENUS.   FLATA, *Fabricius.*
CH. SP. F. exalbida, alis deflexis, elytris punctis marginis interioris antennisque nigris. Expans. alar. fere 2 unc.

F. whitish, with the wings deflexed, the elytra being spotted with black along the posterior margin, antennæ black. Expanse of the wings nearly 2 inches.

SYN.   Flata nigricornis, *Fabr. Syst. Rhyng. p.* 45.
Cicada limbata, var. *Fab. Sp. Ins.* 2. p. 322. *Donovan,* 1st Edit.

This singular insect, and the plant on which it is represented, have an equal claim to attention, both as objects of natural curiosity, and importance in domestic economy. The larva is an elegant and beautiful creature, and China is indebted to its labours for the fine white wax so much esteemed in the East Indies. The plant is not less interesting, as it produces the vegetable tallow, in general use throughout the Chinese empire.

The novelty of these productions could not fail attracting the notice of those learned Europeans who were first permitted to reside in China, and whose object was to promote sciences and arts, as well as the christian knowledge. Both the Wax-insect and Tallow-tree are spoken of in their writings as extraordinary and peculiar advantages to the country. Du Halde, especially in his splendid work l'Histoire de la Chine, treats largely on these productions, in the sections *Cire blanche d'Insectes et l'arbre qui porte le suif.* His relations are, perhaps, too prolix, but they are evidently the result of attentive

Flata nigricornis.

observation, and serve to illustrate the Natural History, and economical purposes of the subjects we are noticing.

The following is the account given by the author: " *De la Cire Blanche, faite par des insectes, et nommée Tchang pe la, c'est-à-dire, Cire blanche d'insectes.*—Ki *dit*. La Cire blanche dont il s'agit ici, n'est pas la même que la cire blanche des Abeilles. Ce sont de petits insectes qui la forment. Ces insectes succent le suc de l'espèce d'arbres nommée *Tong tçin*, et à la longue ils le changent en une sorte de graisse blanche, qu'ils attachent aux branches de l'arbre.

"Il y en a qui disent que c'est la fiente de ces insectes, qui s'attachant à l'arbre, forme cette Cire, mais ils se trompent. On la tire en raclant les branches dans la saison de l'Automne; on la fait fondre sur le feu, et l'ayant passée, on la verse dans l'eau froide où elle se fige, et se forme en pains. Quand on l'a rompue, on voit dans les morceaux brisez, des veines comme dans la pierre blanche ou congélation nommée Pe che cao; elle est polie et brillante: on la mêle avec de l'huile, et on en fait des chandelles. Elle est beaucoup supérieure à celles que font les Abeilles.

" *Chi tchin* dit. Ce n'est que sous la Dynastie des *Yuen* qu'on a commencé à connoître la cire formée par ces insectes. L'usage en est devenu fort commun, soit dans la médecine, soit pour faire des bougies. Il s'en trouve dans les Provinces de *Se tchuen* de *Hou quang*, de *Yunnan*, de *Fo kien*, de *Tche kiang*, de *Kiang nan*, et généralement dans tous les quartiers du Sud-Est. Celle qu'on ramasse dans les Provinces de *Se tchuen* et d'*Yunnan*, et dans les *territoires* de *Hen tcheou*, et de *Yung tcheou* est la meilleure.

" L'arbre qui porte cette cire, a les branches, et les feüilles semblables à celles du *Tong çin*. Il conserve sa verdure durant toutes les saisons: Il pousse des fleurs blanches en bouquets durant la cinquième Lune; il porte des fruits en bayes, gros comme le fruit du *Kin* rampant.

" Quand ils ne sont pas mûrs, ils sont de couleur verte; et ils deviennent noirâtres, lorsqu'ils mûrissent, au lieu que le fruit de *Tong çin* est rouge. Les insectes qui s'y attachent sont fort petits. Quand le soleil parcourt les quinze derniers dégrez des Gémeaux, ils se répandent en grimpant sur les branches de l'arbre; ils en tirent le suc, et jettent par la bouche une certaine bave, qui s'attachant aux branches encore tendres, se changent en une graisse blanche, laquelle se durcit, et prend la forme de cire. On diroit que c'est de la gelée blanche que le froid a durcie.

" Quand le soleil parcourt les quinze premiers dégrez du Signe de la Vierge, on fait la récolte de la Cire, en l'enlevant de dessus les branches. Si l'on diffère à la cueïllir que le Soleil ait entièrement parcouru ce Signe, il est difficile de la détacher, même en la raclant.

" Ces insectes sont blancs quand ils sont jeunes, et c'est alors qu'ils font leur cire.

Quands ils deviennent vieux, ils sont d'un châtain qui tire sur le noir. C'est alors que formant de petits pelotons, ils s'attachent aux branches de l'arbre. Ces pelotons sont au commencement de la grosseur d'un grain de mil: vers l'entrée du printemps ils commencement à grossir et à s'étendre. Ils sont attachez aux branches de l'arbre en forme de grappes, et à les voir, on diroit que l'arbre est chargé de fruits. Quand ils sont sur le point de mettre bas leurs œufs, ils font leur nid de même que les chenilles. Chacun de ce nids ou pelotons contient plusieurs centaines de petits œufs blancs.

"Dans le tems que le soleil parcourt la seconde moitié du Taureau on les cuëille, et les ayant enveloppez dans des feüilles de *Yo* (espèce de simple à larges feüilles); on les suspend à différens arbres. Après que le Soleil est sorti du Signe de Gémaux, ces pelotons s'ouvrent, et les œufs produisent des insectes, qui sortant les uns après les autres des feüilles dont ils sont enveloppez, montent sur l'arbre où ils font ensuite leur cire.

"On doit avoir soin d'entretenir le dessous de l'arbre toûjours propre, et de le garantir des fourmis qui mangent ces insectes. On voit deux autres arbres auxquels on peut attacher les insectes, et qui porteront également de la cire; l'un qui se nomme *Tien tchu*, et l'autre qui est un espèce d'arbre aquatique, dont les feüilles ressemblent assez à celles du Tilleul.

"*Qualitez et effects de cette cire.*—Elle est d'une nature qui n'est ni froide ni chaude, et qui n'a aucune qualité nuisible. Elle fait croître les chairs, elle arrête le sang, elle apaise les douleurs, elle rétablit les forces, elle unit les nerfs, et rejoint les os, prise en poudre dont on forme de pillules, elles fait mourir les vers qui causent la phtisie.

"*Tchi hen* dit. La Cire blanche est sous la dénomination du métal: ses esprits corroborent, fortifient, et sont propres à ramasser et à resserrer. C'est une drogue absolument nécessaire aux chirurgiens: elle a des effects admirables, quand on la fait entrer avec de la peau de *Ho hoang* dans la composition de l'onguent, qui fait renaître et croître les chairs." *Du Halde, Vol. IV. p.* 495, large Folio, 1735.

Sir G. Staunton, in his learned work, has also described the Wax insect; he found it at Turon Bay, in Cochin China, and has caused it to be represented in a vignette plate, with the following description. "Among other objects of natural curiosity, accident led to the observation of some swarms of uncommon insects busily employed upon small branches of a shrub, then neither in fruit nor flower, but in its general habit bearing somewhat the appearance of a privet. These insects, each not much exceeding the size of the domestic fly, were of a curious structure, having pectinated appendages rising in a curve, bending towards the head, not unlike the form of the tail feathers of the common fowl, but in the opposite direction. Every part of the insect was, in colour, of a perfect white, or at least completely covered with a white powder. The particular stem frequented by those insects, was entirely whitened by a substance or powder of that colour, strewed upon it by them. The substance or powder was supposed to form the white wax of the East. This substance is asserted, on the spot, to have the property, by a particular manipulation, of giving in certain proportions, with vegetable oil, such solidity to the composition as to render the whole equally capable of being moulded into candles. The fact is ascertained, indeed, in some degree, by the simple experiment of dissolving one part of this wax in three parts of olive oil made hot. The whole, when cold, will coagulate into a mass, approaching to the firmness of bees' wax."

From the accurate description and figures of the latter author, it is evident, the creature that produces the white wax of China, is an imperfect insect, or technically speaking, the pupa of an insect, which, in its mature state, is furnished with wings. This is clearly the fact, for the rudiments of wings are visible in the figures alluded to.*

Stoll, in his work on *Cimices* and *Cicadæ*, gives a figure of this immature insect under the title of *De Waldraagster* (Nymphe) or *La Cigale Porte Laine*, fig. 144, together with the winged insect at fig. 145; and it is on this authority the latter is introduced in the annexed plate. There is, indeed, much similarity between the pupa and the imago, and some striking characteristics are common to both. They agree in the structure of the antennæ and proboscis, or sucking trunk; the abdomen of the winged insect is also loaded with a fine white powder, and is furnished at the extremity with a tuft of down and hairs, similar to that so eminently conspicuous in the pupa state. We have, however, observed the white powder and tuft on the abdomen of Lystra lanata, and have reason to imagine it also forms a white wax, similar to that of the present species.

Fabricius, in his Species Insectorum, described this insect as a variety of Cicada limbata, which is of a light green colour, with a red margin; that which Stoll has figured, and with which this agrees, is of a pale brown, with a black margin. These are the *species* and *variety* Fabricius describes, for the specimens referred to by Fabricius, in the collection of Sir Joseph Banks, agree precisely with our insects. Fabricius notes the *habitat* Africa. Stoll received the green specimen from the Island of Ceylon; the pale sort from Africa. The larva we have represented is from China; and the imago was brought from the East Indies, by the late Mr. Ellis.

---

* This may account for a passage in *Gordon's* description of China, where he says, " In the plains" of Houquang " are vast numbers of little *worms* that produce wax, in the same manner as bees do honey," if we understand by *worms*, insects not arrived at maturity; for the larva of Bombyx Mori, is also termed a silk worm, though it belongs to the moth tribe when perfect.

HEMIPTERA.

Croton Sebiferum—Poplar-leaved Croton, or Tallow-tree.—The Tallow-tree is not the natural food of the Wax insect, but as they mutually illustrate the same inquiry, they are represented in the same plate; and it is further presumed, that a short account of this useful plant will be deemed a proper sequel to the history of the insect.

Du Halde, when describing the Tallow-tree, says, " Il est de la hauteur d'une grande cerisier. Le fruit est renfermée dans un écorce qu'on appelle *Yen Kiou*, et qui s'ouvre par le milieu quand il est mûr, comme celle de la châtaigne. Il consiste en des grains blancs de la grosseur d'un noisette, dont la chair a les qualitez du suif; aussi en fait-on des chandelles, après l'avoir fait fondre, en y mêlant souvent un peu d'huile ordinaire, et trempant les chandelles dans la cire qui vient sur l'arbre dont je vais parler: il s'en forme autour du suif une espèce de croûte qui l'empêche de couler. Page 18. Vol. I.

Sir G. Staunton speaks nearly to the same effect: " From the fruit of the *Croton sebiferum*, of Linnæus, the Chinese obtain a kind of vegetable fat, with which they make a great proportion of their candles. This fruit, in its external appearance, bears some resemblance to the berries of the ivy. As soon as it is ripe, the capsule opens and divides into two, or, more frequently, three divisions, and falling off discovers as many kernels, each attached by a separate foot-stalk, and covered with a fleshy substance of a snowy whiteness, contrasting beautifully with the leaves of the tree, which, at this season, are of a tint between a purple and a scarlet. The fat, or fleshy substance, is separated from the kernels by crushing and boiling them in water. The candles made of this fat are firmer than those of tallow, as well as free from all offensive odour. They are not, however, equal to those of wax or spermaceti." This author further adds, " The wax for candles is generally the produce of insects, feeding chiefly on the privet, as is mentioned in the chapter of Cochin China. It is naturally white, and so pure as to produce no smoke; but is collected in such small quantities, as to be scarce and dear. Cheap candles are also made of tallow, and even of grease of too little consistence to be used, without the contrivance of being coated with the firmer substance of the tallow-tree or of wax." *Vide Chapter on Sou-choo-foo.*

The tallow-tree is now cultivated in the West Indies, where it thrives well, and produces fruit, and by proper attention may hereafter become useful.

Pl. 18.

Belostoma Indica.

BELOSTOMA INDICA?

Plate 18.

Sub-Order.	HEMIPTERITA, *Kirby*.
Section.	HYDROCORISA, *Latreille*.
Family.	NEPIDÆ, *Leach*.
Genus.	BELOSTOMA, *Latreille*. Nepa p. *Linn*.
Ch. Sp.	B. " squalidè lutea, maculis fuscis, femoribus anticis nigro-lineatis, coxis quatuor posticis immaculatis." Long. Corp. 3 unc. B. dirty clay coloured with brown spots, the anterior femora with black lines, and the four posterior coxæ immaculate. Length 3 inches.
Syn.	Belostoma indica? *Enc. Méth.* X. p. 272. Nepa grandis, *Donovan*, 1st edit. Stoll Cimic. 2. t. 7. f. 4. (Exclus. synon. Linn. Fabr. Merian, Rösel, and De Geer.)

M. Merian has given a plate and description of the South American Belostoma grandis in her work on the Insects of Surinam. We learn from that account, that in the larva and pupa state it lives in the water; that it is a voracious creature, and feeds not only on the weaker kinds of aquatic insects, but on some animals much larger than itself. The pupa* is represented on the back of a large frog in the water, and is designed to portray the manner in which it fastens on those creatures, holding them between its strong curved fore feet, and extracting the juices of their bodies through its singularly constructed beak. M. Merian says, the winged insect was produced from one of these pupæ on the twelfth of May, 1710.

Every writer on this insect since M. Merian appears indebted to her for their account of these few particulars; for though all the European species of the same family undergo precisely the same changes in their aquatic dwellings, among decayed vegetables, &c. at the bottom of the water, and quit it only in the winged state,† we are indebted to her for the time of the appearance of this exotic species in that state, as well as for a correct figure of its pupa.

Linnæus, following Merian, gives *Surinam* as the country of B. grandis; Margravius,

---

* The pupa is *semi-completa*: unlike the pupa of the Lepidoptera, &c. it scarcely differs in appearance or manners of life from the complete insect, but has only the rudiments of the wings. See the lower figure in the accompanying plate.

† Nepa cinerea and linearis are English species of this family; these live in the water till they have wings, when they occasionally quit it to pursue other winged creatures.

HEMIPTERA.

Brasil; and Fabricius, *America* generally. Donovan observed a slight difference between the Chinese specimen and the figures in preceding works referred to by Fabricius; but he nevertheless gave it as the Nepa Grandis of Fabricius, on the authority of the collection of the Right Hon. Sir Joseph Banks, Bart. The Asiatic species are, however, now regarded as specifically distinct from those of America, and hence I have given this doubtingly as the B. indica of Saint Fargeau.

BELOSTOMA (SPHÆRODEMA) RUSTICA.

Plate 19. fig. 1.

Ch. Sp.   B. rotundata, ecaudata, fusca, thoracis elytrorumque margine antico albido.   Long. Corp. lin. $7\frac{1}{2}$.

B. round, without a tail, brown, with the margin of the elytra and the front of the thorax pale. Length of the body $7\frac{1}{2}$ lines.

Syn.   Nepa rustica, *Fabr. Ent. Syst.* 4. 62. 3. (Exclus. syn. *Enc. Méth.* X. p. 273. et *Laporte Revis. Hemipt.* p. 18. Diplonychus rusticus.) *Laporte op. cit. p. 83.* Nepa plana, *Sulz. Hist. Ins. t.* 10. *f.* 2.   *Stoll Cim.* 2. *t.* 7. *f.* 6.

Insects in general discover an extraordinary degree of care and ingenuity in depositing their eggs in the most secure situations, or places where the infant brood, when hatched, may be provided with proper sustenance. Those of the aquatic kind usually lay them in recesses in the mud or sand, or under loose stones that lie at the bottom of the water: others, with as much care, and more ingenuity, hollow out the interior substance of the large stalks of water plants, and deposit their eggs in them; or, rising out of the water, lay them in the extreme branches of those plants, to secure them from other aquatic depredators. Belostoma rustica displays even more sagacity, or attachment for its eggs, than those creatures; for it never leaves them. Till they are hatched, it bears them on its back, in a cluster of an oval shape; these eggs are of an oblong form, and are fastened by the narrowest end to a thin film, or plate of cement, that causes them to adhere to the polished surface of the wing cases; when these eggs, about a hundred in number, are hatched, it casts off the exuviæ of the cluster, and differs no longer in general appearance from the male of the same species.

Our figures represent the situation of the eggs on the back, and the insect also after they are cast off. It is not commonly received with the eggs upon it. Found on the coast of Coromandel, as well as China.

1. Belostoma rustica.   2. Nepa rubra.

1. *Velledia Dufrai*   2. *Nanbuschii*

HEMIPTERA.

NEPA RUBRA.

Plate 19. fig. 2.

GENUS. NEPA, *Linnæus.*
CH. SP. N. oblongo-ovata, depressa, fusca, longè bicaudata, abdomine supra rubro, linea nigrâ. Long. Corp. (cauda exclus.) 1¼ unc.
N. oblong-ovate, depressed, brown, with two long filaments at the anus, abdomen above red, with a black dorsal line. Length 1⅓ inch.
SYN. Nepa rubra, *Linn. Syst. Nat.* 2. 713. 2. *Fabr. Syst. Rhyng. p.* 107.

CALLIDEA OCELLATA.

Plate 20. fig. 1.

SECTION. GEOCORISA, *Latreille.*
FAMILY. SCUTELLERIDÆ, *Leach.*
GENUS. CALLIDEA, *Laporte.* Chrysocoris, *Hahn.* Cimex p. *Linn.* Tetyra p. *Fabricius.*
CH. SP. C. carnea thorace scutelloque maculis flavescentibus, quibusdam puncto ocellari atro. Long. Corp. ¾ unc.
C. orange-red, thorax and scutellum with yellow spots, some of which have a black central spot. Length ¾ inch.
SYN. Cimex ocellatus, *Thunberg Nov. Sp. f.* 72.
Tetyra dispar, *Fab. Ent. Syst.* 4. *p.* 81. *Syst. Rhyng. p.* 129. *Stoll Cimic. t.* 37. *f.* 260. *Donov.* 1st edit.

This curious insect is among the number of those lately brought from China. A figure of the upper surface is represented on a leaf of the *Camellia Sesanqua*, one of the vignette plates of Sir G. Staunton's History of the late Embassy to that country; a coloured figure and a short account of it may therefore prove acceptable to the readers of his volumes.

Stoll has also given a figure of it, and has also represented another sort, which he considers as the female *(letter A)*; it has no black points in the yellow spots of the thorax and scutellum: he mentions the Isle of Formosa as the native country of his specimens.

Fabricius observes, that it varies according to the sex by having an acute spine on each side of the thorax, as represented in the accompanying figures; this statement is, however, denied in the Encyclopédie Méthodique, vol. x. p. 410, by Messrs. Saint Fargeau and Serville. The absence of these spines, although not perhaps a sexual character, certainly indicates a remarkable variety. I have certainly seen specimens destitute of these spines.

The pupa figured at the bottom of the plate on the left hand side is probably that of Tetyra Druræi or some allied species. Donovan is entirely silent respecting it.

ASTEMMA SCHLANBUSCHII.

Plate 20. fig. 2.

FAMILY.	LYGÆIDÆ.
GENUS.	ASTEMMA, *Laporte.* Lygæus, *Fabr.*
CH. SP.	A. sanguinea, thorace fasciâ abbreviatâ, scutello, elytrorum puncto alisque atris. Long. Corp. fere ¾ unc.
	A. red, thorax with an anterior abbreviated black band; scutellum, wings, and a spot on each wing-cover, black. Length nearly ¾ inch.
SYN.	Lygæus Schlanbuschii, *Fabr. Ent. Syst.* 4. *p.* 155. *Syst. Rh. p.* 222.

The accompanying figure does not represent the anterior transverse thoracic black band mentioned in the Fabrician description; the species varies in this respect. The pupa figured on the right hand side at the bottom of the plate is most probably that of this insect, although Donovan is entirely silent concerning it.

CALLIDEA STOCKERUS.

Plate 21. fig. 1.

FAMILY.	SCUTELLERIDÆ.
GENUS.	CALLIDEA, *Laporte.* Cimex p. *Linn.* Tetyra p. *Fabr.*
CH. SP.	C. cærulea, thorace punctis 6, sc. 3 parvis anticis, 3 majoribus posticis; scutello 7, apiceque nigris. Long. Corp. lin. 6.
	C. blue, the thorax with 6 spots, three small in front, and three behind larger; the scutellum with 7 large spots, and the apex black. Length 6 lines.
SYN.	Cimex Stockerus, *Linn. Syst. Nat. p.* 715. *Fabr. Ent. Syst.* 4. *p.* 79.
	Cimex Stollii, *Wolff.*

This insect seems to be very common in China; for we rarely receive a parcel of the insects of China that does not include many of them. There are several distinct, but very

1. Callidea Stockerus. 2. Raphigaster aurantius.
3. ─── cruciger. 4. ─── Phasianus.
5. ─── bifidus.

closely allied, species, which have been regarded as varieties, but which appear to be constant in their variations, being inhabitants likewise of distinct countries. These vary not only in their tints and the number of their spots, but also in their outline, which seems clearly to prove them specifically distinct. Donovan mentions one of these as " a charming miniature variety of C. Stockerus from *Africa æquin.* about one-third of the size of the Chinese specimens, and of a very deep blue colour. The marks both on the upper and under side precisely resemble those in the annexed figures."

RAPHIGASTER AURANTIUS.

Plate 21. fig. 2.

FAMILY.	PENTATOMIDÆ.
GENUS.	RAPHIGASTER, *Laporte.* Edessa p. *Fabr.*
CH. SP.	R. aurantius, capite thoracis margine antico, abdominis maculis marginalibus pedibusque atris. Long. Corp. 1¼ unc.
	R. orange-coloured, with the head, anterior margin of the thorax, lateral spots of the abdomen, antennæ and legs, black. Length 1¼ inch.
SYN.	Cimex aurantius, *Fabr. Ent. Syst.* 4. *p.* 105. *Syst. Rhyng. p.* 149. (Edessa a.) *Sulzer Ins. t.* 10. *f.* 10. *Stoll Cimic, t.* 6. *f.* 39.

CHARIESTERUS CRUCIGER.

Plate 21. fig. 3.

FAMILY.	COREIDÆ, *Leach.*
GENUS.	CHARIESTERUS, *Laporte, Burm.* Lygæus p. *Fabr.*
CH. SP.	Ch. thorace acutè spinoso; oblongus, supra niger, thorace lineis, elytris cruce, ferrugineis. Long. Corp. lin. 10.
	Ch. with the angles of the thorax acutely spined; oblong, black above, thorax with two lines and the margins orange, elytra with an orange cross. Length 10 lines.
SYN.	Cimex cruciger, *Fabr. Mant. Ins.* 2. 289. 104.
	Lygæus cruciger, *Fabr. Ent. Syst.* 4. *p.* 141.

Donovan states, that this insect was " from the collection of Mr. Francillon, who received it from China. Fabricius describes it as a native of Brazil." This is certainly the true locality of the species. Donovan's figure was taken from a specimen which wanted the terminal joint of the antennæ.

CERBUS TENEBROSUS?

Plate 21. fig. 4.

GENUS.   CERBUS, *Hahn.* Anisoscelis, *Latr.* Lygæus p. *Fabr.*
CH. SP.   C. cinnamomeo-fuscus, antennis apice tibiisque dilutioribus.   Long. Corp. unc. 1.
          C. cinnamon-brown, with the tips of the antennæ and tibiæ paler.   Length 1 inch.
SYN.   Lygæus tenebrosus? *Fabr. Ent. Syst.* 4. p. 135.   *Burm. Handb. der Ent.* 2. p. 340.
          Lygæus Phasianus, *Donov.* 1*st edit.*

Donovan gave this as identical with the Lygæus Phasianus, Fabr. from tropical Africa, although his specimens were brought from China by the late Mr. Ellis. This figure appears rather to represent a female of the common Lyg. tenebrosus.

HARPACTOR BIFIDUS.

Plate 21. fig. 5.

FAMILY.   REDUVIIDÆ.
GENUS.   HARPACTOR, *Laporte.* Reduvius p. *Fabr.*
CH. SP.   H. ater, elytris fasciâ rufâ, scutello spinâ erectâ apice bifidâ.   Long. Corp. $\frac{9}{10}$ unc.
          H. black, elytra with a red bar, scutellum with an erect spine forked at the extremity. Length $\frac{9}{10}$ of an inch.
SYN.   Riduvius bifidus, *Fabricius Ent. Syst.* 4. 204.   *Syst. Rhyng.* p. 285.
          Cimex bifidus, *Donovan,* 1*st edit.*

Pl. 22.

Papilio Paris.

LEPIDOPTERA.

Order. **LEPIDOPTERA.** *Linnæus.*

PAPILIO PARIS.

Plate 22.

TRIBE.   Diurna, *Latreille.* (Papilio, *Linnæus.*)
FAMILY.   Papilionidæ, *Leach.*
GENUS.   Papilio, *Linnæus*, (Section, Equites.) *Latreille, Boisduval, &c.*
CH. SP.   P. alis nigris, aureo-viridi pulverulentis, posticis caudatis, maculâ (in utroque sexu) magna azureo-cæruleâ, ocello fulvo ad angulum ani, his subtus maculis septem marginalibus ocellatis. Expans. alar. 4 unc.

P. with the wings black, powdered with golden-green atoms, the posterior with a broad tail and a large shining blue spot in both sexes, and a reddish eye at the anal angle; beneath with seven marginal eye-like spots. Expanse of the wings about 4 inches.

SYN.   Papilio Paris, *Linn. Syst. Nat.* 2. p. 745. No. 3. *Fabr. Ent. Syst.* 3. 1. p. 1. No. 1. *Drury Exotic Ins.* V. I. t. 12. f. 1. 2. *Cramer Pap.* 2. pl. 103. f. A. B. *Esper. Ausl. Schmett.* t. 2. f. 1. *Encycl. Méth.* IX. p. 69. *Boisduval Hist. Nat. Lep.* 1. p. 208.

The simile proposed by Linnæus, both for the arrangement and specific nomenclature of butterflies, is gleaned from ancient and fabulous history. The species are divided into sections of Trojan and Greek princes, heroes, deities, nymphs, and plebeians: and the species have received names in accordance with this fanciful theory, which, at least, in the writings of Linnæus is well conducted, and seems liable to less objection than the characters assigned to each section: for many species placed among the Equites, and a more considerable number with the Plebeii, are inconsistent with the essential criterion Linnæus has given. This arrangement has necessarily undergone material alterations in the *Entomologia Systematica* of Fabricius and other still more recent works; alterations certainly justified by the more comprehensive views now taken of this pleasing branch of Entomology. The Equites of Fabricius, with many additions, and a few exceptions, are the same as those in the two Linnæan sections: Papilio Priamus is, however, removed from the head of the *Equites Trojani*, and the precedence given to Papilio Paris.

Papilio Paris is an insect of considerable beauty. The general colour on the upper surface is obscure brown, nearly approaching black, but finely contrasted with brilliant green atoms, profusely sprinkled over it. The posterior wings are adorned with a large blue spot, which derives additional lustre from the dusky colour surrounding it. Another species, very similar to Papilio Paris, but without this spot, is also found in China. It has been supposed to be the female of our species, which opinion is adopted in the Encyclopédie Méthodique. Fabricius names it Bianor, after *Cramer*, *pl.* 103. *fig.* 6. Dr. Horsfield has described and figured another species, from Java, in his Lepidoptera Javanica, under the name of Papilio Arjuna, but which is so closely allied to Paris, that it may eventually prove to be only a geographical variety. Its larva is cylindrical, with a coriaceous shield-like plate, extending over the three anterior segments of the body; the chrysalis is greatly angulated, with the head notched. Another species, or at least strong variety, has been lately received from the Himalayan mountains.

PAPILIO CRINO.

Plate 23.

Ch. Sp. P. alis nigris atomis viridi-aureis, fasciâ communi cæruleo-viridi; posticis caudatis, ocello anali rufo, his subtus lunulis viridibus cæruleis cinereisque. Expans. alar. $3\frac{1}{2}$ unc.

P. with the wings black and sprinkled with golden-green atoms, with a greenish-blue bar running across all the wings, the posterior pair tailed, with a red eyelet at the anal angle; beneath with green, blue, and ashy lunules. Expanse of the wings $3\frac{1}{2}$ inches.

Syn. Papilio Crino, *Jones. Fabricius Ent. Syst.* 3. 1. *p.* 5. *Enc. Méth.* IX. *p.* 66. *Boisduval Hist. Nat. Lepid.* 1. *p.* 207.
Papilio Regulus, *Stoll Suppl. Cramer.* 5. *pl.* 41. *f.* 1.

This splendid butterfly is extremely rare, and its precise country is doubtful. Fabricius says, "Habitat in *Africa*. Mus. Dom. Drury." Donovan, however, who had access to Drury's collections, says, "We have found an unique specimen of this species in the collection of Mr. Drury, and on that authority we include it as a native of China. Fabricius erroneously gives Africa as its locality." In the Encyclopédie Méthodique, Africa is given. Boisduval gives "Indes orientales;" his unique specimen having been sent to him by M. Drege as from Cochin China, but which Boisduval thinks may possibly be erroneous. The manuscripts of Drury, now in my possession, throw no light upon the subject further than that there are several *unnamed* species indicated as inhabitants of China as well as of Sierra Leone. But from the strong affinity between Crino, Palinurus, Paris, &c. it is scarcely to be doubted that China or India is the real locality of Crino.

Renealmia exaltata, a majestic plant, near seven feet in height, bearing a fine pendant group of flowers at the summit, is figured in the plate.

Papilio Crino.

Pl. 24.

1. Papilio Ascon.   2. Papilio Agenor.

LEPIDOPTERA.

PAPILIO COON.

Plate 24. fig. 1.

Ch. Sp.
P. alis angustatis, anticis elongato-ovatis fuscis; posticis caudâ spatuliformi, atris, maculis baseos palmatis, lunulis submarginalibus albis maculâque duplici ad angulum ani flavà. Expans. alarum 4½—5¼ unc.

P. with narrow wings, the anterior elongate ovate, brown on both sides; the posterior, with a spatulate tail, black, with palmated basal spots and submarginal lunules of a white colour; and two yellow spots at the anal angle. Expans. of the wings 4½—5¼ inches.

Syn.
Papilio Coon, *Jones. Fabricius Ent. Syst.* 3. 1. *p.* 10. *Enc. Méth.* IX. *p.* 65. *Boisduval Hist. Nat. Lep. p.* 201.
Papilio Hypenor, *Enc. Méth.* IX. *p.* 65.

The original Fabrician description was derived from a specimen in the collection of Mr. Drury, and Donovan's figure is copied from the drawings of Mr. Jones, referred to by Fabricius. The translation of the Fabrician description of the lower wings is incorrectly rendered in the Encyclopédie Méthodique, and in consequence another description is given of a Javanese specimen of this species, under the name of P. Hypenor. It has recently been received in considerable numbers from Java, from whence I possess a specimen with the wings much longer and narrower than they are here represented.

PAPILIO AGENOR.

Plate 24. fig. 2.

Ch. Sp.
P. alis nigris, basi sanguineis, anticis striatis, posticis dentatis, disco albo maculisque marginalibus atris. Expans. alar. 6 unc.

P. with the wings black, bloody at the base, the anterior with longitudinal paler markings, the posterior dentate with a white disc and black marginal spots. Expansion of the wings 6 inches.

Syn.
Papilio Agenor, *Linn. Syst. Nat.* 2. *p.* 747. *Fabr. Ent. Syst.* 3. 1. *p.* 13. *Enc. Méth.* IX. *p.* 28. *Clerk Ic. t.* 15. *Cramer, pl.* 32. A. B. *Herbst. Pap. t.* 8. *f.* 3.
Papilio Memnon, ♀ *Boisduval Hist. Nat. Lepid. p.* 193.

This is one of the largest Chinese Papiliones we are acquainted with. The upper and under surfaces so nearly agree, that Donovan considered a figure of the first unnecessary. M. Boisduval has advanced several forcible reasons for regarding this and

LEPIDOPTERA.

several other allied insects (P. Laomedon, of Cramer, Anceus, Achates) as females of the very variable Asiatic species, Papilio Memnon; the caterpillar of which, according to Dr. Horsfield, is green, with the anterior segments narrowed and retractile, the third being elevated at the top and marked with an eye-like spot on each side. It feeds upon the species of Citrus.

The plant figured is *Plumbago Rosea. Rose-coloured Lead-wort.*

PAPILIO PERANTHUS.

Plate 25.

Ch. Sp. P. alis nigris, supra basi cærulescenti-viridibus, subtus apice pallidis, posticis obtuse dentatis, caudatis, subtus lunulis rufescentibus serie digestis. Expans. alar. 4 unc.

P. with the wings black, greenish blue at the base; beneath paler at the external margins, posterior pair dentate and tailed, with red lunules on the under side arranged in a transverse series. Expansion 4 inches.

Syn. Papilio Peranthus, *Fabricius Ent. Syst.* 3. 1. p. 15. *Enc. Méth.* IX. p. 66. *Boisduval Hist. Nat. Lepid.* 1. p. 203.

The original Fabrician description was made from a specimen from Cochin China in the Banksian collection. Donovan mentions another which came from Canton, and M. Boisduval gives Borneo, Java, and Celebes, as its localities.

The insect is represented on a small twig of *Arundo Bambos (Bamboo* or *Cane)*, a well known plant, mentioned by Sir G. Staunton as being one of the most useful productions of China.

PAPILIO TELAMON.

Plate 26. fig. 1.

Ch. Sp. P. alis caudatis, concoloribus, flavescentibus, maculis fasciisque nigris, posticis utrinque strigâ sanguineâ nigromarginatâ. Expans. alar. 3 unc.

P. with the wings coloured alike pale yellowish, with black spots and bands, the posterior with very long narrow tails, and a red streak bordered with black at the anal angle. Expansion of the wings 3 inches.

Syn. Papilio Telamon, *Donovan. Boisduval Hist. Nat. Lep.* 1. p. 250.

The singular delicacy and beauty of this Papilio is not the only claim it has to the particular attention of Entomologists: it is clearly an undescribed species; and perhaps

Pl. 25

Papilio Peranthus.

1. Papilio Telamon. 2. Papilio Agamemnon.

the only specimen of it yet brought to Europe is that from which our figure is copied. It was taken near Pekin, by a gentleman in the suite of Earl Macartney, in the embassy to China; and was originally in the possession of Mr. Francillon, of London, who kindly permitted drawings and descriptions to be made of this and every other insect in his magnificent collection that could enhance the value of this publication. It is still so rare that M. Boisduval states that he had never seen a specimen of it.

Papilio Telamon bears a distant resemblance to P. Protesilaus, but a much stronger to P. Ajax: pursuing then the metaphorical method of arranging the butterflies in the Linnæan manner, the name of the father of Ajax, who was one of the distinguished Grecian Princes at the siege of Troy, has been given to this species.

PAPILIO AGAMEMNON.

Plate 26. fig. 2.

Ch. Sp. P. alis nigris viridi-maculatis, posticis breviter caudatis, his subtus ocello lunato maculisque rubris. Expans. alar. 3½ unc.

P. with the wings black and spotted with pale green, the posterior pair with short tails and ornamented beneath with a lunate eyelet and red spots. Expansion of the wings 3½ inches.

Syn. Papilio Agamemnon, *Linn. Syst. Nat.* 2. *p.* 748. *Fabr. Ent. Syst.* 3. 1. *p.* 33. *Enc. Méth.* IX. p. 46. *Boisduval Lep.* 1. p. 230.
Papilio Ægistus, Cramer, 106. C. D. (corrected, p. 151.)

Papilio Agamemnon is found in several parts of Asia (China, Bengal, Java, the Moluccas and Philippine Islands, Manilla, Timor). The under side is beautifully adorned with a number of bright green spots of various sizes. The general colour is pale pink, diversified with shades of chestnut brown. The upper side is much plainer; the general colour is black, except the spots, which are green, and precisely agree in shape with those on the under side. Dr. Horsfield has figured the transformations of this insect (Lepid. Javan. pl. 4. f. 12), the larva is short and thick, with a forked tail; the chrysalis has the head very obtuse. This species is the type of Dr. Horsfield's second section of the genus, having the club of the antenna oval and compressed.

PAPILIO PROTENOR. ♀

Plate 27.

Ch. Sp. P. alis anticis fuscis nigro-striatis; posticis dentatis, nigris, atomis pallidis, maculâ duplici rufâ anguli ani.   Expans. alar. 5½ unc.
P. with the anterior wings brown with black longitudinal stripes, the posterior dentate, black with pale atoms, and a double red spot at the anal angle.   Expans. of the wings 5½ inches.

Syn. Papilio Protenor, *Fabricius Ent. Syst.* 3. 1. 13. *Cramer, pl.* 49. A. B. *Enc. Méthod.* IX. *p.* 30. *Boisduval Hist. Nat. Lepid. p.* 198.
♀ Papilio Laomedon, *Jones. Fabricius Ent. Syst.* 3. 1. 12. *Donovan, 1st edition*, (*nec Cramer, pl.* 50. *f.* A. B.)

The collection of the late Mr. Latham contained the original specimen from which Mr. Jones' drawing (referred to by Fabricius) was made. The present figure was copied, by Mr. Jones' permission, from that drawing.

PAPILIO EPIUS.

Plate 28. fig. 1.

Ch. Sp. P. alis nigris, flavo maculatis; posticis dentatis fasciâ irregulari, maculis adjectis flavis, maculâque anguli analis rufâ.   Expans. alar. fere 3⅛ unc.
P. with the wings black, spotted with yellow; the posterior pair dentate with an irregular yellow bar, accompanied externally with additional spots and a large red spot at the anal angle.   Expanse of the wings nearly 3½ inches.

Syn. Papilio Epius, *Jones. Fabricius Ent. Syst.* 3. 1. *p.* 35. *Enc. Méth.* IX. *p.* 43. *Boisduval Hist. Nat. Lepid. p.* 238.
Papilio Erithonius, *Cramer, pl.* 232. A. B.
Papilio Demoleus, *Esper. Ausl. Schmett, tab.* 50. *f.* 1—4.

Papilio Epius and Papilio Demoleus are so similar in their marks and colours, that many authors have confounded one species with the other. Papilio Epius is chiefly distinguished by the red spot in the interior margin of the lower wings, having no blue eye-shaped mark above it.

Papilio Protenor.

1. Papilio Epius.   2. Papilio Demoleus.

Morpho Rhetenor.

LEPIDOPTERA.

PAPILIO DEMOLEUS.

Plate 28. fig. 2.

Ch. Sp. P. alis nigris, flavo-maculatis, posticis dentatis fasciâ flavâ subrectâ ocelloque anali dimidiatim cæruleo rufoque. Expans. alar. 3½ unc.

P. with black wings spotted with yellow, the posterior dentate with a nearly straight and regular yellow fascia, and an ocellus at the anal angle blue above and red beneath. Expanse of the wings 3½ inches.

Syn. Papilio Demoleus, *Linn. Syst. Nat.* 2. 753. *Fabr. Ent. Syst.* 3. 1. 34. *Kleeman (Rosel, add.) t.* 1. *f.* 2. 3. *Cramer Ins. t.* 231. *f.* A. B. *Boisduval Hist. Nat. Lep.* 1. *p.* 237. *Encycl. Méth.* IX. *p.* 43.
Papilio Demodocus, *Esper. Ausl. Schmett. t.* 51. *f.* 1.

Linnæus gives the Cape of Good Hope as the habitat of this species; and Boisduval also mentions the coast of Guinea, Senegal, and Madagascar. Fabricius, however, particularly says, " Habitat in Indiæ orientalis Citro, Dr. Koenig," describing the larva as solitary, smooth, of a yellowish green colour, with a reddish head, two tentacles on the neck, and a bifid tail. Boisduval has, however, applied this observation to P. Epius, stating that P. Demoleus had been reared at Senegal by M. Dumolin, and that its larva feeds on the citron trees.

MORPHO RHETENOR.

Plate 29.

Family. Nymphalidæ, *Swainson.*
Genus. Morpho, *Fabricius (Syst. Gloss. in Illig. Mag.)*
Ch. Sp. M. alis suprà nitidissimè cyaneis; subtùs umbrino griseoque variis, ocellis cæcis. Expans. alar. 5½ unc.

M. with the wings on the upper side dazzling cyaneous blue, beneath varied with umber and grey, with blind eyelets. Expanse of the wings about 5½ inches.

Syn. Papilio Rhetenor, *Cramer, pl.* 15. A. B. *Herbst. Pap. t.* 27. *f.* 1. 2. *Esper. Pap. Exot. t.* 42. *f.* 1. *Sulzer Ins. t.* 13. *f.* 1. *Enc. Méth.* IX. *p.* 444.

Whatever effect the artist can produce by a combination of the most brilliant colours employed in painting, must be far surpassed by comparison with the dazzling appearance of this splendid creature. It is impossible to find in any part of the animal creation colours more beautiful or changeable. Pale blue is the principal colour, but new tints meet the eye in every direction, varying from a silvery green to the deepest purple; and the whole surface glittering with the resplendence of highly polished metal.

This splendid species was long confused with Papilio Menelaus of Linnæus, the authors of the Encyclopédie Méthodique, however, cleared up the confusion, proving them to be quite distinct. The species, and indeed the entire group to which it belongs, are, however, natives of South America. Sulzer, indeed, states that his specimen came from China, and evidently on this authority Donovan introduced the species into this work.

Thea Laxa (Bohea, or broad-leaved Tea), is figured in the plate. Sir G. Staunton says, the bohea tea is supplied in China from the province of Fochen: the green tea from Kiang-nan. The leaves of these teas vary in some degree in form according to the age of the plant; those of the bohea are the broadest; Thea stricta has much longer leaves, they are lanceolated, and more deeply serrated than those of the bohea. Many authors have considered them varieties of the same species. It flowers in England in August and September.

ACRÆA VESTA.

Plate 30. fig. 1.

FAMILY. HELICONIIDÆ, *Swainson.*
GENUS. ACRÆA, *Fabricius.* (Heliconia p. *Fabricius olim.*)
CH. SP. A. alis oblongis integerrimis, utrinque corticinis; omnium supra limbo posteriori fusco serieque punctorum interrupto. Expans. alar. $2\frac{1}{2}$ unc.
A. with the wings oblong and entire, of a pale yellow brown; with a dark brown border in which are white spots. Expanse of the wings $2\frac{1}{2}$ inches.
SYN. Papilio (Helic.) Vesta, *Jones. Fabr. Ent. Syst.* 3. 1. p. 163. *Enc. Méth.* IX. p. 233.
Papilio Terpsichore, *Cramer Pap. pl.* 298. *f.* A. B. C.

Papilio Vesta is the only insect of the *Heliconii* division of Butterflies described by Fabricius as peculiar to China, in his *Ent. Syst.*, the majority being inhabitants of Africa. It is a rare species. The *Papilio Vesta* of Cramer is a very different insect, being the *P. Erato* of Fabricius.

1. Acraea Vesta. 2. Pieris Pasithoe. **
3. Pieris Hyparete.

PIERIS PASITHOE.

Plate 30. fig. 2.

FAMILY. PAPILIONIDÆ, *Leach*.
GENUS. PIERIS, *Fabricius, Boisduval* (Papilio, Heliconia, *Linn.*)
CH. SP. P. alis suboblongis, nigris, supra cærulescenti-albo-maculatis; posticis subtus disco flavo, nigro venoso, fasciàque basali ferruginea. Expans. alar. 3½ unc.
P. with oblong wings of a black colour spotted on the upper side with bluish white, the posterior pair with the disc beneath yellow, with black veins and a broad red basal fascia. Expanse of the wings 3½ inches.
SYN. Papilio (Helicon.) Pasithoe, *Linn. Syst. Nat.* 2. 755. *Fab. Ent. Syst.* 3. 1. 179.
Pieris P. *Encycl. Méth.* IX. *p.* 148. *Boisduval Hist. Nat. Lep.* 1. *p.* 451. *Drury Illust. Exot. Ent.* 2nd edit. *v.* 2. *p.* 16.
Papilio (Dan. Cand.) Dione, *Drury*. 1st edit.
Papilio Porsenna, *Cramer, pl.* 43. *fig.* D. E. *and pl.* 352. *fig.* A. B.

PIERIS HYPARETE.

Plate 30. fig. 3.

CH. SP. P. alis suboblongis, integerrimis, albis, utrinque apice, subtus venis nigris, posticis subtus, plus minusve flavis, maculis sanguineis limbo nigro apicali digestis. Expans. alar. 3 unc.
P. with the wings rather oblong, entire, white with a black margin on both sides and with black veins beneath; the posterior on the under side more or less stained with yellow, with a row of six red spots in the black border. Expanse of the wings about 3 inches.
SYN. Papilio (Helicon.) Hyparete, *Linn. Syst. Nat.* 2. *p.* 763. (nec *Fabr. Ent. Syst.* 3. 1. *p.* 176. *Enc. Méth.* IX. *p.* 153. *Boisduval Hist. Nat. Lep.* 1. *p.* 455.
Papilio Autonoe, *Cramer*, 187. C. D. et 320. A. B.

Several distinct, but nearly allied, species have been confounded together under the name of Hyparete; Donovan observed, " We have two sorts of this species; one with the marginal row of red spots on the posterior wings disposed in a deep border of black; the other has the red spots on a whitish ground. They are certainly the two sexes of Papilio Hyparete. Found near Canton, in China." The sexes do not, however, vary in this respect, the latter individuals mentioned in this passage are therefore most probably the

P. Eucharis of Drury (Epicharis, Enc. Méth. and Boisduval; Hyparete, Fabricius), or the P. Autonoe of Cramer and Boisduval.

The leaf represented in the plate is that of *Sophora Japonica (Shining-leaved Sophora)*, an elegant and valuable timber tree, of which Sir G. Staunton speaks as very frequent in China.

PIERIS (IPHIAS) GLAUCIPPE.

Plate 31. fig. 1.

Sub-Gen. Iphias, *Boisduval.*

Ch. Sp. P. alis supra albis, anticis maculâ magnâ apicali (medio fulvo) nigrâ, subtus (nisi dimidio basali anticarum) cinereis, strigis minutis fuscis irroratis. Expans. alar. 4 unc.

P. with the upper sides of the wings white, the anterior having a large black spot at tip, the middle of which is rich orange; beneath, except the basal half of the anterior wings, greyish with brown waves. Expanse of the wings 4 inches.

Syn. Papilio (Dan. Cand.) Glaucippe, *Linn. Syst. Nat.* 2. 762. *Fabr. Ent. Syst.* 3. 1. p. 198. *No.* 618. *Herbst. Pap. Fab.* 96. *f.* 1—3. *Enc. Méth.* IX. p. 119. *Drury Exot. Lepid.* 1. *pl.* 10. *f.* 3. 4. *2nd edit.* p. 20. *Horsfield Lep. Jav.* p. 130. (Colias Gl.)

♀ Pap. Callirhoe, *Fab. Mant. Ins.* 2. 20. 215.

Iphias Glaucippe, *Boisduval Hist. Nat. Lep.* 1. p. 596.

P. Glaucippe is an elegant insect: very common in China, and it is said, in some adjacent parts of Asia, also Bengal, Java, &c. The Papilio Callirhoe of Linnæus is considered as the female of this species: few authors deem it more than a variety (β). Dr. Horsfield has described and figured the caterpillar and chrysalis of this species in his work upon the Lepidoptera of Java above referred.

1. Pieris Glaucippe. 2. Pieris Pyrene

1. Colias Pyranthe.   2. Colias Philea.

LEPIDOPTERA.

PIERIS (THESTIAS) PYRENE?

Plate 31. fig. 2

Sub-Gen. THESTIAS, *Boisduval.*
Ch. Sp.   P. " alis integerrimis, rotundatis, flavis; primoribus apice (medio fulvo) nigris; subtus nebuloso-maculatis." *Linn. loc cit. infra.* Expans. alar. fere 2½ unc.
         P. with the wings entire, rounded, yellow; the anterior with a large black spot, the centre of which is bright orange; beneath with cloud-like marks. Expanse of the wings about 2½ inches.
Syn.   Papilio (Dan. Cand.) Pyrene, *Linn. Syst. Nat.* 2. 762. 86? *Enc. Méth.* IX. *p.* 120. Boisduval *Hist. Nat. Lep.* 1. 593. (Thestias P.) *Drury Exot. Ent.* 2nd edit. 1. *p.* 11. *pl.* 5. *f.* 2.
         Papilio Sesia, *Fabricius Ent. Syst.* 3. 1. *p.* 203. *Donovan, 1st. edit.*

The insect here figured, judging at least from the upper side, agrees with the Linnæan description of P. Pyrene, the habitat of which is given by Linnæus as China. Fabricius, however, who refers to Linnæus, gives America as its locality. It is now satisfactorily ascertained that it inhabits China and various parts of the East Indies.

     The plant figured is *Limodorum Tankervillæ*, an elegant and much admired production of China.

COLIAS (CALLIDRYAS) PYRANTHE.

Plate 32. fig. 1.

Genus.   COLIAS, *Fabricius.* Papilio (Danai Candidi), *Linnæus.*
Sub-Gen. CALLIDRYAS, *Boisduval.*
Ch. Sp.   C. alis integerrimis, rotundatis, albis, puncto apiceque nigris; subtus cinereo undulatis puncto fulvo. Expans. alar. fere 3 unc.
         C. with the wings entire, rounded, and white, having a discoidal spot and the margin black; beneath with ashy waves and a fulvous spot. Expanse of the wings nearly 3 inches.
Syn.   Papilio Pyranthe, *Linn. Syst. Nat.* 2. *p.* 763. *Enc. Méth.* IX. *p.* 97. *Boisduval Hist. Nat. Lep.* 1. *p.* 611.

     Papilio Gnoma of Fabricius; P. Alcyone, Cramer; P. Nepthe, Fabricius; and P. Chryseis, Drury (1st edition); are in all probability varieties of this insect.

COLIAS (CALLIDRYAS) PHILEA.

Plate 32. fig. 2.

Ch. Sp. C. alis integerrimis, subangulatis, flavis; anticis macula, posticis limbo, luteis. Expans. alar. 3¾ unc.

C. with the wings entire, somewhat angulated, bright yellow, with a large discoidal spot in the anterior and a broad margin of the posterior pair orange. Expanse of the wings 3¾ inches.

Syn. Papilio (Dan. Cand.) Philea, *Linn. Syst. Nat.* 2. 764. *Fabricius Ent. Syst.* 3. 1. *p.* 212. *Cramer Pap. pl.* 173. E. F. *Roesel Ins. Bel. t.* 3. *f.* 5. *Boisduval Hist. Nat. Lep.* 1. *p.* 619. (Callidryas P.)

Linnæus says of this species, "Habitat in Indiis," which may be either taken for the East or West Indies; Roesel calls it "die *indianische* goldborte;" and Donovan states that his specimen was received from China. The real locality, however, not only of this butterfly, but of all the species of the section to which it belongs, is South America and the West Indies. M. Boisduval gives the Argante of Hubner, Lolia of Godart, Aricia of Cramer, Melanippe Cramer, and Larra of Fabricius, as varieties of the female of this species.

The plant represented in the plate is *Melastoma Chinensis*.

MORPHO (DRUSILLA) JAIRUS.

Plate 33.

Family. Nymphalidæ, *Swainson*.

Genus. Morpho, *Fabricius*. (Papilio, *Dan. Festiv. Fabr. olim.*)

Sub-Gen. Drusilla, *Swainson Zool. Illust.* 1. *pl.* 11. (Hyades, *Boisduval Hist. Nat. Lep.* 1. *pl.* 13. *f.* 1.)

Ch. Sp. M. alis integris, fuscis; posticis disco baseos albo, supra oculo maximo, subtus duobus dissitis. Expans. alar. 4½ unc.

M. with the wings entire, brown; the posterior, with the basal disc white, with a single large eye on the upper and two on the lower side. Expanse of the wings 4½ inches.

Syn. Papilio F. Jairus, *Fabr. Ent. Syst.* 3. 1. *p.* 54. *Cramer Pap. pl.* 6. A. B. and *pl.* 185. A. B. C. *Enc. Méth.* IX. *p.* 445.

Papilio Cassiæ, *Clerk Icon. tab.* 29. *fig.* 3.

A specimen of this extremely rare Butterfly was contained in the collection of Dr. Hunter, now the property of the University of Glasgow; a fragment in the British Mu-

Morpho Jairus.

Pl. 34

Nymphalis Bernardus.

LEPIDOPTERA.

seum; and one in fine preservation in the collection of Mr. Francillon. Except these, and the specimens from which the figures in the annexed plate are copied, Donovan had never seen it in any cabinet whatever. It had been figured only by two authors, Clerk in his *Icones insectorum rariorum*, and Cramer in his *Papillons exotiques*. The figures of Clerk and Cramer do not strictly agree: we observe those of the first much lighter coloured, and the white space on the upper wings considerably larger than in any of the figures in Cramer's plates.

Fabricius says it is a native of the East Indies. One specimen figured by Cramer was brought from the isle of Amboyna. It seems therefore not peculiar, like some insects, to China.

NYMPHALIS (CHARAXES) BERNARDUS.

Plate 34.

GENUS. NYMPHALIS, *Latreille*. (Papilio Nymphalis, *Fabricius*.)
SUB-GEN. CHARAXES, *Boisduval*. (Jasia, *Swainson*)
CH. SP. N. alis fulvis, anticis apice atris, fasciâ mediâ flavâ, posticis caudatis strigâ punctorum ocellatorum. Expans. alar. $3\frac{1}{2}$ unc.
N. with fulvous wings, the anterior black at the tips with a broad pale yellow band, the posterior tailed, with a row of black ocellated spots. Expanse of the wings $3\frac{1}{2}$ inches.
SYN. Papilio (Nymph.) Bernardus, *Jones. Fabricius Ent. Syst.* 3. 1. *p.* 71.

This uncommonly rare Chinese butterfly has not been figured in any other work. Fabricius described it only from the drawings of Mr. Jones. I possess a specimen in which the central fascia is nearly white, and is continued half way across the posterior wings, and the black spots in the latter are very broad and confluent, without white in the centre.

The plant represented is the *Camellia Japonica (Japan Rose)*, a native of Japan and China, which blossoms from January to May. It is a lofty and magnificent plant, rising to the height of several feet: there is a variety of it with double flowers, perfectly white; and another in which the flowers are variegated with white and red.

LEPIDOPTERA.

ARGYNNIS ERYMANTHIS.

Plate 35. fig. 1.

Genus. Argynnis, *Fabricius.*

Ch. Sp. A. alis subrotundatis, subdentatis, fulvis, anticis fasciâ flavescenti transversâ mediâ nigro-maculatâ, apice nigris; posticis serie punctorum duabusque lunularum nigrarum. Exp. alar. 2—3 unc.

A. with the wings rather rounded and indented, fulvous, the anterior with a transverse pale yellow fascia, spotted with black, the tips black; the posterior wings with a row of black spots and two rows of narrow spots. Expanse of the wings from 2 to 3 inches.

Syn. Papilio (Dan. Fest.) Erymanthis, *Drury Exot. Ent. vol.* 1. *pl.* 15. *f.* 3. 4. *Cramer, pl.* 238. *f.* 9. *Fabr. Ent. Syst.* 3. 1. *p.* 139. *Enc. Méth.* IX. *p.* 257.

Papilio Lampetia, *Cramer Pap. pl.* 148. *fig.* E.

It is the rarity, and not the beauty of this butterfly, which induced Donovan to add it to this selection. It is probably far from common in China, being very seldom sent to Europe among the insects of that country.

CYNTHIA ORITHYA.

Plate 35. fig. 2.

Genus. Cynthia, *Fabricius.* (Papilio Nymphales gemmati, *Linn.*)

Ch. Sp. C. alis denticulatis, supra nigris aut fuscis, singularum ocellis duobus iride fulvâ; anticis costâ strigisque apicalibus albis, his subfalcatis; posticis rotundatis. Expans. alar. 2. unc.

C. with the wings dentate, the anterior subfalcate, the posterior rounded; above blacker brown shaded with blue, each having two eyelets with a fulvous circle; the anterior margin and several apical fasciæ white. Expanse of the wings 2 inches.

Syn. Papilio N. Orithya, *Linn. Syst. Nat.* 2. *p.* 770. (nec *Abbot* and *Smith Lep. Georg.* v. 1. *t.* 8.) *Roesel Ins.* 4. *t.* 6. *f.* 2. *Cramer Pap. pl.* 19. C. D. 32. E. F. 281. E. F. 290. A. B. C. D.

Papilio N. Orythia, *Fabr. Ent. Syst.* 3. 1. *p.* 91. *Donovan,* 1st edit.

Donovan observes, that "the varieties of Papilio Orythia are numerous, and seem to differ according to climate of the countries of which they are natives. It is common in North America, Jamaica, India, &c. The variety from North America is almost wholly

1. Argynnis Erymanthis. 2. Cynthia Orithya.
3. Limenitis Leucothoe. 4. Limenitis Eurynome.

brown, and those from Jamaica have less blue in the disk of the lower wings than those from China." Donovan, however, clearly here mistook distinct species for varieties. The American species thus named by Abbot and Smith is the P. Larinia of Fabricius, and in my copy of the Entomologia Systematica, which belonged to Professor Weber, the companion of Fabricius, the words "India orientali" are introduced in lieu of Jamaica.

Papilio Clelia of Cramer, which is found on the coast of Guinea, has been supposed a variety of Papilio Orythia. Fabricius, in the *Entomologia Systematica*, has made it a distinct species. It greatly resembles P. Orythia, but has no more blue colour on the posterior wings than is concentrated in a large spot near the base.

LIMENITIS LEUCOTHOE.

Plate 35. fig. 3.

GENUS. LIMENITIS, *Fabricius.* (Papilio Nymphales Phalerati, *Linn.*)
CH. SP. L. alis dentatis, supra fusco-nigris, subtus fulvis; utrinque fasciis tribus macularibus albis; posticarum fasciâ intermediâ punctis nigris antrorsum fœtis. Expans. alar. 2½ unc.
L. with the wings dentate, above brownish black, beneath clay coloured; with three rows of white spots on both sides, the intermediate fascia of the posterior wings with black spots towards the base. Expanse of the wings 2½ inches.
SYN. Papilio Leucothoe, *Linn. Syst. Nat.* 2. p. 780. *Fabr. Ent. Syst.* 3. 1 p. 129. *Enc. Méth.* IX. p. 430. *Herbst. Pap.* tab. 240. *f.* 5. 6.
Papilio Polyxina, *Donov.* 1st edition.

Donovan regarded this as a new species, giving the following as the true Leucothoe of Linnæus. The description given by that author, and especially his notice of the third row of spots in the posterior pair of wings being composed "ex maculis 7 albis puncto nigro fœtis," clearly applies to this and not to the following insect.

K

LIMENITIS EURYNOME.

Plate 35. fig. 4.

Ch. Sp.   L. alis dentatis supra fusco-nigris, subtus fulvis; fasciis interruptis macularibus albis, subtus fusco cinctis, anticis fasciâ longitudinali baseos è maculis duabus triangularibus compositâ. Expans. alar. 2½ unc.

L. with the wings dentate, above brownish black, beneath fulvous, with interrupted white maculated bands, which on the under side are edged with brown, the anterior have also a longitudinal basal fascia, composed of two triangular white spots, the bases of which are opposed to each other. Expanse of the wings 2½ inches.

Syn.   Limenitis Eurynome, *Westw.*
Papilio Leucothoe, *Donov. 1st edit.*
Papilio Aceris major ex India, *Esper. Pap. tab.* 82. *f.* 1.

CYNTHIA ŒNONE.

Plate 36. fig. 1.

Ch. Sp.   C. alis denticulatis supra luteis margine omni nigro; posticis basi late nigris, maculâ cyaneâ. Expans. alar. 2½ unc.

C. with the wings denticulated, above pale clay-coloured, with all the margins black, the base of the posterior black with a large cyaneous blue spot. Expanse of the wings 2½ inches.

Syn.   Papilio (Nymph. Gemm.) Œnone, *Linn. Syst. Nat.* 2. 770. " alis denticulatis, primoribus albido maculatis subbiocellatis, posticis basi cyaneis ocellis duobus."
Fabr. Ent. Syst. 3. 1. 90. *Kleman Ins.* 1. *t.* 3. *f.* 1. 2.
Vanessa Œnone, *Enc. Méth.* IX. p. 318.

Donovan says, that this insect is found throughout Asia (which is the locality assigned to it by Linnæus and Fabricius), and is very common in China. In the Encyclopédie Méthodique, the Cape of Good Hope is given as its habitat. The Linnæan specific character is applicable to the female; the male (according to M. Godart), which is here figured, having no eyes on the upper side of the wings.

1. Cynthia Oenone.　　2. Cynthia Almana.
3. Nymphalis Lubentina.

LEPIDOPTERA.

CYNTHIA ALMANA.

Plate 36. fig. 2.

Ch. Sp.   C. alis anticis falcatis, posticis intus subcaudatis, omnibus supra fulvis, ocellis sesquialtero, subtus fuscescentibus, posticis lineâ flavidâ transversâ mediâ. Expans. alar. 2¾ unc.

C. with the anterior wings falcate, the posterior subcaudate at the inner angle, all fulvous above, with an ocellus on each; beneath brownish, the posterior with a yellowish transverse line in the centre. Expanse of the wings 2¾ inches.

Syn.   Papilio (N. G.) Almana, *Linn. Syst. Nat.* 2. 769. *Fabr. Ent. Syst.* 3. 1. p. 89. *Cramer Pap. pl.* 58. F. G. *Herbst. Pap. t.* 172. 1. 2.

The angulated form of the wings of this butterfly gives it a remarkable appearance. The eyes on the wings somewhat resemble those of the Peacock butterfly, to which, in some other respects, it bears no distant similitude. It is common in China; Fabricius gives its *habitat* Asia.

NYMPHALIS (ACONTHEA) LUBENTINA.

Plate 36. fig. 3.

Ch. Sp.   N. alis subdentatis, fusco-virescentibus; anticis utrinque fasciâ albâ, maculari; posticis apice punctis chermisinis, serie duplici digestis. Expans. alar. 2½ unc.

N. with the wings subdentate, brownish-green, the anterior on each side with a row of white spots, the posterior with scarlet spots arranged in a double series towards the extremity. Expanse of the wings 2½ inches.

Syn.   Papilio (Nymph.) Lubentina, *Fabricius Ent. Syst.* 3. 1. p. 121. *Enc. Méth.* IX. 400. *Cramer Pap. pl.* 155. C. D. *Herbst. Pap. t.* 146. 1. 2.

Aconthea Lubentina, *Horsfield Lep. Jav. pl.* 5. *f.* 5.

Papilio Lubentina is figured only in the works of Cramer: his specimen is not precisely like ours, but agrees in all the essential peculiarities, and is unquestionably the same species. The semitransparent spots on the anterior wings are much larger in Cramer's figure than in the insect before us.

LEPIDOPTERA.

NYMPHALIS JACINTHA.

Plate 37. fig. 1.

Ch. Sp.   N. alis repando-dentatis, fuscis; anticis striga punctorum alborum, posticis apice albis margine fusco lunulis albis. Expans. alar. 4—4½ unc.

N. with the wings scalloped, brown; the anterior with a row of spots on the anterior pair at the tips, posterior externally white, the margin being brown with white lunules. Expanse of the wings from 4 to 4½ inches.

Syn.   Papilio (Nymph. Phal.) Jacintha, *Drury, app. vol.* 2. *pl.* 21. *f.* 1. 2. *Fabricius Ent. Syst.* 3. 1. *p.* 60. ♀
Papilio (Nymph.) Liria, *Fab. Ent. Syst.* 3. 1. *p.* 126. ♂?
Papilio Perimale, *Cramer, pl.* 65. C. D. 67. B.

This curious butterfly was found in the province of Pe-tche-lee, in China. It is in all probability the female of Pap. (N.) Liria, Fabr.

It is represented, with P. Antiochus, on a leaf of the *Urtica Nivea (White Nettle).**

NYMPHALIS ANTIOCHUS.

Plate 37. fig. 2.

Ch. Sp.   N. alis supra holosericeo-nigris, fasciâ communi nitide aurantiâ; anticarum abbreviatâ. Expans. alar. 2¾ unc.

N. with the wings above holosericeous black, with a broad shining orange bar common to all the wings, but abbreviated in the anterior pair. Expanse of the wings 2¾ inches.

Syn.   Papilio (Dan. Fest.) Antiochus, *Linn. Mant.* 1. 537. *Drury, app. vol.* 3. *pl.* 7. *f.* 3. 4. *Fabricius Ent. Syst.* 3. 1. *p.* 44. *Enc. Méth.* IX. *p.* 409.
Papilio Eupalemon, *Cramer, tab.* 143. *f.* B. C. Le Velouté. *Daubenton, pl. enl.* 68. *f.* 3. 4.

This insect is very rare in European cabinets of insects. The specimen figured by Drury came from the Brazils, and Cramer's from Surinam. Fabricius, however, describes

---

* Sir G. Staunton speaks of a cloth that the Chinese manufacture from the fibres of a dead nettle. Query, Is this the species employed for that purpose? no other is noticed by that author in the lists of plants collected in China. The nettle is of general use in Russian Tartary also; the *Kuriles*, and other Siberian tribes, make cloth, cordage, thread, &c. of it. *Gordon*, &c.

1. Nymphalis Jacintha. 2. Nymphalis Antiochus.

Nymphalis Sylla.

LEPIDOPTERA.

it as a native of China; and Donovan states that the insect figured in the collection of drawings of Mr. Jones, of Chelsea, was a native of China, as well as the specimen in his own collection. There must, however, have been some mistake in respect to these specimens, for not only are all the immediately allied species natives of South America, but Stoll observed its transformations in that country, and says, that the caterpillar feeds on the tamarind; it is green, with two long spines on the head, and numerous other shorter spines on the body.

NYMPHALIS SYLLA.

Plate 38.

Ch. Sp. N. alis dentatis, supra nigris viridi-maculatis striatisque; anticis fasciâ maculari niveâ. Expans. alar. 3¾ unc.

N. with the wings dentate, black on the upper side, with green spots and lines, the anterior with a row of white spots, the central fascia of the posterior wings externally radiated. Expanse of the wings 3¾ inches.

Syn. Papilio Sylla, *Cramer Pap. t.* 43. *f.* F. G.
Papilio Sylvia, *Herbst. Pap. t.* 247. *f.* 2. 3.
Papilio (N.) Gambrisius, *Fabr. Ent. Syst.* 3. 1. *p.* 85. *Donov. 1st edit.*
Nymphalis Sylvina, *Enc. Méth.* IX. *p.* 381.

A specimen of this very rare Papilio was taken in one of the small islands on the eastern coast of China, and was in the possession of Mr. Francillon. Sir J. Banks, Bart. had a specimen of it from another part of the East Indies. It also occurs in Java and Amboyna (Enc. Méth.).

MYRINA (LOXURA) ATYMNUS.

Plate 39. fig. 1.

FAMILY. LYCÆNIDÆ, *Swainson*.
GENUS. MYRINA, *Fabr. Enc. Méth.*
SUB-GEN. LOXURA, *Horsfield Lep. Jav. p.* 119.
CH. SP. M. alis supra fulvo-testaceis, apice nigro, posticis longè caudatis. Expans. alar. $1\frac{1}{2}$—$1\frac{3}{4}$ unc.
M. with the wings above orange red, with the tips black, the posterior with very long tails white at the tips. Expanse of the wings $1\frac{1}{2}$—$1\frac{3}{4}$ inches.
SYN. Papilio (Pleb. Rural.) Atymnus, *Fab. Ent. Syst.* 3. 1. *p.* 283. *Cramer, pl.* 331. *fig.* D. E. (palpis deteritis.)
Loxura Atymnus, *Horsfield Lep. Jav. p.* 119. *pl.* 2. *f.* 6. *Boisduval Hist. Nat. Lep.* 1. *pl.* 7. *fig.* 3.

This is also a scarce species. Donovan's specimen was from the collection of the late Duchess Dowager of Portland, who procured it from China. Another specimen in the cabinet of Sir J. Banks, Bart. is from Siam. Dr. Horsfield found it in Java.

The plant represented in the plate is *Hemerocallis Japonica*, brought from China by Mr. Slater.

THECLA MÆCENAS.

Plate 39. fig 2.

GENUS. THECLA, *Fabricius*. (Papilio Hesperia Pleb. rural., *Fabricius olim.*)
CH. SP. Th. alis bicaudatis atris disco cæruleo, subtus brunneo nebulosis. Expans. alar. $1\frac{3}{4}$ unc.
Th. with the wings black and furnished with two tails, the disc being blue; beneath clouded with brown. Expanse of the wings $1\frac{3}{4}$ inches.
SYN. Hesperia (R.) Mæcenas, *Jones. Fab. Ent. Syst.* 3. 1. *p.* 271. 45. *Enc. Méth.* IX. *p.* 639.

1. Myrina Atymnus. 2. Thecla Mæcenas.

1. Deilephila Nechus. 2. Glaucopis Polymena.

LEPIDOPTERA.

DEILEPHILA NECHUS?

Plate 40. fig. 1.

Section. Crepuscularia, *Latreille.*
Family. Sphingidæ, *Leach.*
Genus. Deilephila, *Ochsenh.* Sphinx p. *Linn. &c.*
Ch. Sp. D. " alis integris; anticis viridibus; strigâ testaceâ; posticis nigris; maculis baseos fasciaque flavis." *Fabric. loc. cit. subtus.* Expans. alar. 3½ unc.
D. with the wings entire, the anterior green with a testaceous streak, (" with testaceous marks," *Donov.*) the posterior black with spots at the base, and a row of spots near the extremity. Expanse of the wings 3½ inches.
Syn. Sphinx Nechus? *Fabricius Ent. Syst.* 3. 1. p. 377. *Cramer Ins. t.* 178. f. B.

The number of Chinese species of this genus, already described, is very limited: the insect represented in the accompanying figures is the largest of them; but as this is inferior in size to several kinds found in Europe, we conceive there must remain many larger species of the genus unknown to collectors of foreign insects, and yet very common in China. In the latter part of Sir G. Staunton's work, that author mentions the larva of a *Sphinx Moth* which furnish an article for the table of the Chinese. We regret that the indefinite expression cannot assist us to determine the species, and scarcely the genus, of the insect alluded to.*

The specimen figured in the annexed plate was in the collection of Mr. Francillon, who received it from China. The habitat of D. Nechus, given by Fabricius, is America: and Cramer has represented a small variety of the same species from North America.

---

* European naturalists are entirely ignorant of the Chinese insects in the state of larva and pupa, if we except a few species of the Cimices, Cicada, and some altogether uninteresting insects, that have been accidentally brought among others from that country. Hence it must remain undetermined whether they correspond in form with those of other parts of the world. It is, however, highly probable, from their great affinity to those in the perfect state, that in the state of larva they may also agree. The extensive collection of the larvæ of sphinges made by Mr. Abbot in North America affords no singularly constructed animal distinct from those found in Europe; they vary indeed in their colours, but preserve uniformly the characters found on the same genus in other countries. We noticed among the drawings of the late Mr. Bradshaw the figure of a Chinese sphinx, apparently *S. Hylas*, together with a larva similar to that of the *S. Stellatarum*: it was green, and, like all the known larvæ of the family (except the *Adscitæ* division), was perfectly free from hairs: it was also furnished with a horn at the posterior part of the body.

LEPIDOPTERA.

Moreover, Donovan's figure and description do not precisely correspond with the Fabrician description, so that on both these grounds I have considered it advisable to give the specific name with a mark of doubt. Sphinx Batus and Sphinx Gnoma are nearly allied to this insect, particularly the former; both are found in different parts of the East Indies.

GLAUCOPIS POLYMENA.

Plate 40. fig. 2.

FAMILY.   ZYGÆNIDÆ, *Leach.*
GENUS.   GLAUCOPIS, *Fabr.* Sphinx, *Linn. Donov.*
CH. SP.   G. nigra alis maculis luteis, anticis tribus, posticis duabus; abdomine cingulis duobus coccineis. Expans. alar. fere 2 unc.
          G. black, wings spotted with deep yellow, the anterior having three and the posterior two spots, abdomen with two scarlet bands. Expanse of the wings nearly 2 inches.
SYN.   Sphinx Polymena, *Linn. Syst. Nat.* 2. 806. *no.* 40. *Cram. Ins. t.* 13. *f.* D. *Fabr. Ent. Syst.* 3. 1. *p.* 396. *Drury Exot. Ent.* 1. *t.* 26. *f.* 1.

This beautiful creature is probably scarce in China; at least it is very rarely found among the insects brought from that country.

It is figured on the plate with the *Rosa semperflorens (Ever-blowing China Rose)*.

SESIA HYLAS.

Plate 41. fig. 1.

GENUS.   SESIA, *Fabricius.* Sphinx, *Linn. Donov.*
CH. SP.   S. alis fenestratis, abdomine barbato viridi, cingulo purpureo. Expans. alar. 2½ unc.
          S. with transparent wings, body pale yellow green, abdomen with a brush at the tip and a purple belt round the middle. Expanse of the wings 2½ inches.
SYN.   Sphinx Hylas, *Linn. Mant.* 1. 539. *Fabr. Ent. Syst.* 3. 1. *p.* 379.
          Sphinx Picus, *Cramer Ins. t.* 148. *f.* B.

1. Sesia Hylas. 2. Callimorpha? Thallo.
3. Callimorpha? ruficollis. 4. Callimorpha? bifasciata.

LEPIDOPTERA.

CALLIMORPHA? THALLO.

Plate 41. fig. 2.

SECTION. NOCTURNA?
FAMILY. ARCTIIDÆ, *Stephens*?
GENUS. CALLIMORPHA? *Latreille.* Sphinx, *Linn.*
CH. SP. C. alis oblongis integerrimis nigris anticis fasciis duabus, posticis unica flavis; capite rubro. Expans. alar. 2 unc.
C. with oblong entire black wings, anterior pair shaded with blue at the base, and with two pale yellow fasciæ; posterior wings with a pale yellowish space, head red. Expanse of the wings 2 inches.
SYN. Papilio Thallo, *Linn. Syst. Nat.* 2. 756. *Fabr. Ent. Syst.* 3. 1. *p.* 173.
Sphinx pectinicornis, *Linn. Syst. Nat.* 2. 807. *Fab. Ent. Syst.* 3. 1. *p.* 399. *Edwards Aves*, 36. *t.* 226.
Phalæna tiberina, *Cramer, t.* 32. *f.* C. D.
Sphinx Thallo, *Donov.* 1*st edit.*

Donovan entered into a lengthened observation, shewing that Fabricius had given a Papilio Thallo in all his works when no such Papilio was in existence, and that Edwards' figure of the insect in question was derived from a mutilated or mended specimen. The former error is, however, rather to be attributed to Linnæus, who introduced all the confusion by describing Edwards' figure both as a Papilio and Sphinx.

The figures of Cramer and Edwards do not precisely agree; in the former, the disk of the posterior wings is yellowish, with a deep border of black: in the other, the yellow occupies only a space near the base, and forms a semi-lunar mark near the anterior margin of those wings. Donovan suspected, with Cramer, that they are but the two sexes of one species. Cramer says both his specimens came from China, from whence Donovan's insects were also received. The real affinities of this and some other allied species are very perplexing, they seem, however, to connect the Zygænidæ with the Arctiidæ.

CALLIMORPHA? RUFICOLLIS.

Plate 41. fig. 3.

Ch. Sp. C. " alis integerrimis nigro-purpurascentibus, fasciâ communi maculisque duabus flavis, thorace antice brunneo." Expans. alar. 1¼ unc.

C. with the wings entire, black purple, a semicircular yellowish band communicating across all the wings, and two spots of the same colour near the apex, collar reddish. Expanse of the wings 1¼ inch.

Syn. Sphinx ruficollis, *Donov. 1st edit.*

This and the following species were considered by Donovan to be undoubted nondescripts: both species were in the collection of Mr. Francillon, who received them from China.

CALLIMORPHA? BIFASCIATA.

Plate 41. fig. 4.

Ch. Sp. C. alis fulvis; anticarum fasciâ apiceque nigris. Expans. alar. 1 unc.

C. with orange or fulvous wings, anterior pair with a black bar across the middle, and the tips black. Expanse of the wings 1 inch.

The plant represented is *Thuja Orientalis (China Abor-vitæ Tree)*, an ornamental evergreen, much esteemed by the Chinese, and very frequently represented in their landscapes. Sir G. Staunton remarks, in the account of the journey from Pekin to Canton, that great quantities of this plant grew to a prodigious height in the valley in which stands the city of Yen-choo-foo.

Saturnia Atlas.

LEPIDOPTERA.

SATURNIA ATLAS.

Plate 42.

FAMILY. BOMBYCIDÆ.
GENUS. SATURNIA, *Schranck.* (Phalæna Attacus, *Linn.*)
CH. SP. S. alis anticis falcatis, luteo variis, macula fenestrata anticis sesquialtera. Expans. alar. 8 unc.

S. with the anterior wings falcate; yellow brown, varied with paler markings, each wing with a triangular talc-like spot in the middle, the anterior having also a smaller one near the tips. Expanse of the wings 8 inches.

SYN. Phalæna Attacus Atlas, *Linn. Syst. Nat.* 2. *p.* 808. *Petiv. Gaz. t.* 8. *f.* 7. *Fab. Ent. Syst.* III. 1. *p.* 407. *Cram. Ins.* 381. C. 382. A.

The nocturnal Lepidoptera are remarkable for the neatness and simplicity of their colours. Their elegancies consist in the infinite variety and delicacy of intermingled tints: the contrast of spots, specklings, and lineations, which constitute the minutiæ of insect beauty. Some species are to be excepted in this remark; the larger kinds are often gaudy, and the smallest exhibit a display of the richest colours, fancifully disposed, and most elegantly diversified.

The European species are numerous, and pretty well ascertained: those of remote countries remained, at the period of the publication of the first edition of this work, and still remain in great obscurity. The species inhabiting China are almost unknown;* for Fabricius describes not more than twenty species in all the cabinets in Europe. From this scanty number a few are selected to illustrate the genus, and if these appear deficient in point of interest or variety, it may stimulate others to collect new species whenever an opportunity occurs. The moths, not only of China, but of every country except Europe, have received but little attention. In Europe, the number of this tribe exceeds that of any other: on the contrary, the extra European species are comparatively the most inconsiderable of our acquisitions. The Papiliones, or butterflies, are a showy and lively race: they sport in the open fields by day, and attract the traveller's curiosity; hence our cabinets abound with them. But the moths, infinitely more numerous, and not less pleasing, are seldom seen; in the gloominess of their dispositions, they seek the

---

* *Fab. Ent. Syst.* These are chiefly described from insects in the collection of "*Monson Londini,*" of which no figures are extant, and the collection unknown.

LEPIDOPTERA.

obscurity of the forest in the day, and only venture on the wing when the sun is down. In Europe we visit their nocturnal haunts without difficulty or dread; but in hotter climates these are oftentimes impenetrable, or the lurking places of ferocious animals; and few will expose themselves to their attacks to increase the catalogue of exotic moths.*

Phalæna Atlas is the first species we have to notice. It is one of the largest of the moth tribe,† and is, indeed, a gigantic creature. The species is common, but not peculiar to China, being found in other parts of Asia, and in America. The influence of climate is easily traced in the individuals from different countries; that from Surinam is the largest, and of the deepest colours. The Chinese kind is the next in size; the colours incline to orange, and the anterior wings are more falcated, or hooked, at the ends. We

---

* The far greater number of moths can only be taken in the woods at night. This is termed *mothing* by collectors. The moths begin to stir about twilight, and when almost dark, commence their flight. The collector is furnished with a large gauze folding-net, in which the insects are caught indiscriminately, for it is impossible to distinguish one species from another, and often is so dark, that the object itself can barely be discerned. Different species have their favourite haunts, some the lanes and skirts of woods, but many of them prefer the open breaks in the most retired places. As it would be unsafe, or impossible, to penetrate the woods in many countries, it is better to collect the larvæ, or caterpillars, for these may be found on the trees in the day-time, and if kept in little gauze cages, and carefully fed, will change into chrysalides, and produce the moths. This is certainly tedious, and few travellers will divert their attention from more important observations; but were they to appropriate their leisure to this branch of science, they would materially improve entomology. Mr. Abbot investigated a small district of Georgia, in North America, in this manner, and our cabinets are indebted to his labours for several hundred species, altogether new in Europe. The reader may estimate the importance of these discoveries, by referring to the two expensive volumes of North American Lepidopterous Insects; and reflecting, that the originals of all the species included in that work are but a small selection from those he has furnished us with. Viewing these as the result of one man's research, in an inconsiderable portion of North America, what a variety of new and splendid kinds would be the reward of those, who should explore the more genial regions of Asia, Africa, and South America, with equal diligence and information!

We have hazarded an assertion which may seem inadmissible, that the Phalænæ are infinitely more numerous than the Papiliones, or any other tribe of insects. Not that we possess more, but because, in every country that has been investigated, experience justifies such opinion. For instance, in Great Britain we have only sixty[a] Papiliones, and by mere accident two or three local species have lately been added; of the Phalænæ we have more than 900. The same comparative proportion is observed throughout the countries of the European continent; and it is singularly analogous, that our opinion is confirmed, by the recent discoveries of Mr. Abbot in America.

† When Linnæus described it, few of the very large species of Phalæna were known. There are two species from the interior of Africa, which are larger than the Chinese Atlas, and several others scarcely inferior in magnitude.

---

[a] (There are now about eighty-five indigenous British butterflies, and between seventeen and eighteen hundred moths. J. O. W.)

have two other Asiatic *varieties* still smaller, with the wings extremely falcated. These are to be regarded as distinct species.

The larva of Phalæna Atlas is figured by *M. Merian*, in the *Insecta Surinamensia*, plate 52: it is about four inches in length, green, with a yellow stripe disposed longitudinally. Upon each segment are four distinct round tubercles, of a coral-like orange colour, which are surrounded with very delicate hairs. The pupa is large, and inclosed in a web of an ochre colour. The silk of this web is of a strong texture, and it has been imagined, that if woven, it would be superior in durability to that of the common silk worm. *Seba* has also represented the larva at fig. 1. plate 57. vol. 4. *Thesaurus Naturæ*. It is nearly six inches in length, and bulky in proportion; the Phalæna is also larger than that figured by Merian, which is a small specimen of the Surinam species. According to Merian, there are three broods of this insect in a year; they are very common, and feed on the orange trees. Linnæus says, they adhere so tenaciously to the leaves that they can scarcely be taken off.*

The common silk worm, or Phalæna Mori, belongs to this family, and merits observation as a native of China. The art of weaving its threads into silk is of the earliest date. The discovery is attributed to the *Seres*, a people of the East Indies, supposed the Chinese.† In the days of Solomon, we are told, a woman named Pamphilia, of the Island of Co, was skilled in the art of making cloth of the silk brought from the country of the Seres. The most ancient of the Chinese writers ascribe the invention to one of the women of the emperor *Hoang ti*, named *Si ling*, and in honour *Yuen fei*.‡ When Rome degenerated into voluptuousness, Persia, its dependency, furnished this article of luxury; but it is supposed they were indebted to the Chinese for it, and being supplied only in small quantities, it was consequently dear. In Rome it was so scarce, as to be worn only by persons of the first distinction.

The Chinese historians affirm, that the discovery was considered at first of such importance, that all the women in the palace of the emperor were engaged in rearing the insect and weaving its silk. In after times, the silk of China was a principal article of commerce; but latterly, its value has been materially lessened by the culture and fabrication of silk in other countries. As the Chinese know little of the use of linen, the silk is a staple article of their own consumption. The jesuit missionaries mention several

---

* Larva verticillata verrucis pilosis nec folliculos grandes, tenaces, vix extricandos. *Linn. Syst. Nat.*

† Velleraque ut foliis depectant tenuia Seres. *Virg. Georg.* II. 122.

‡ Du Halde, *Des Soyeries*. Les plus anciens écrivans de cet empire en attribuent la decouverte à une des femmes de l'Empereur *Hoang ti*, nommée *Si ling*, et surnommée par honneur *Yuen fei*.

sorts of it in use among the Chinese; some admired for beauty, and others for durability. It is generally supposed these are not merely the effect of different manufacture, but are the produce of distinct insects.* Sir G. Staunton speaks of the culture of silk worms

---

* M. Merian says, in the description of the Surinam variety of Phalæna Atlas: " Telam ducunt fortem, quare bonum fore sericum rata, istius aliquam collegi copiam et in Belgium transmisi, ubi eadem optima judicata est: ut itaque, si quis Erucas istas congregandi laborem non detrectaverit, et bonæ notæ bombycem, et maximum hinc lucrum sibi comparare posset." The thread of which this coccon's web is composed is so strong, that it has been imagined it would make good silk. I have brought some of it into Holland, which has been esteemed such; so that if any one would take the trouble to collect a number of these caterpillars, they would be found good silk worms, and produce great profit. *Merian.*—Abbot informs us, the Moths of the Emperor tribe in general are called silk worms by the people of Georgia; and in the description of Phalæna Cecropia is still more explicit: for he says, " the caterpillar spins on a twig; the outside web is coarse, the inner covered with silk, like a silk worm's coccon. It is said this silk has been carded, spun, and made into stockings, and that it will wash like linen." *Abbot's Ins. by Dr. J. E. Smith.*—These insects are all of the same natural order, P. Cecropia is rather smaller, but very similar to P. Atlas, and this information at least corroborates the assertion of Merian.

An opinion that the Chinese rear several kinds of insects for the sake of their silk has long been prevalent. Dr. Lettsom proposes a query on this subject, " Which species of moth or butterfly is it, the caterpillar of which, in China, affords that strong grey kind of silk, and how is it manufactured or wore? How are these silk worms or caterpillars preserved, fed, and managed? The introduction of such a new silk into England would be a useful acquisition, and redeem entomology from the censure it is now branded with, of being a mere curiosity void of any real utility." [a] If *Lesser* and *Lyonet* are to be relied on, the *Théoloyie des Insectes* answers this query. " At this day there are to be found in China, in the province of Canton, silk worms in a wild state, which, without any care being taken of them, make in the woods a kind of silk which the inhabitants afterwards gather from the trees. It is grey, without lustre, and is used to make a very thick and strong cloth, named there Kien Tcheon. It may be washed like linen cloth, and does not stain." A gentleman resident in the East Indies speaks of a large Phalæna producing silk in that country: " We have a beautiful silk worm north-east of Bengal, that feeds on the Ricinus, whence I call it Phalæna Ricini; it is sea-green, with soft spines, very large and voracious, and spins a coarse, but strong and useful silk. The moth is of great size, with elegant dark plumage. Is it known to European naturalists?" *In a collection of papers published by Dr. Anderson in Madras,* 1788, 1789.—M. Le Bon, Reaumur, Roesel, and several others, have attempted to weave the silk of spiders as a substitute for that of silk worms, but their experiments rather amuse and point out the ingenuity of the proposers than promise to be useful; for after many trials, it appears that the silk of spiders would be inferior in lustre and far more expensive than that of silk worms. Sir G. Staunton alludes to these experiments in his description of the Java forests. " In some open spots were found webs of spiders, woven with threads of so strong a texture, as not easily to be divided without a cutting instrument; they seemed to render feasible the idea of him who, in the southern provinces of Europe, proposed a manufacture from spiders' threads, which is so ridiculous to the eyes of those who have only viewed the flimsy webs such insects spin in England." Many other substances of a soft texture have also been wrought into a variety of trifling articles, as gloves, stockings, &c. of the fibres of Asbestos earth, or mountain flax, beard of the large *Pinna* shell, &c. &c.

---

[a] Naturalist and Traveller's Companion, 1774.

1. Heleona milataris.   2. Eusemia lectrix.

LEPIDOPTERA.

in China, but only of the common sort. It will gratify curiosity, if not prove advantageous, should future observers ascertain what kind of insects the Chinese appropriate to making silk, and whether P. Atlas is of the number, as has been conjectured. It is indeed to be observed, that in India several distinct species of Saturnia are known to be employed in the production of silk; the most important of which are the Tusseh (S. Paphia, Linn.), the Arrindi (S. Cynthia, Drury), and the Kolisurra silk worm of the Dukhun.*

HELEONA MILITARIS.

Plate 43. fig. 1.

FAMILY. ARCTIIDÆ?
GENUS. HELEONA, *Swains. Zool. Illustr. N. Ser.* 116.
CH. SP. H. alis patulis concoloribus luteis apice maculisque violaceis, anticis extus albo-maculatis. Expans. alar. 3½ unc.
H. with the wings extended at rest, the anterior and posterior pairs coloured alike, luteous yellow, with the extremity and spots at the base violet, the anterior with whitish spots at the tips. Expanse of the wings 3½ inches.
SYN. Phalæna militaris, *Linn. Syst. Nat.* 2. 811. *Fabricius Ent. Syst.* 3. 2. p. 416. *Roesel. Ins.* 4. *t.* 6. *f.* 3. *Cramer. Ins. t.* 29. *f.* B.

The natural situation of this and some other allied insects is doubtful; it forms the type of Mr. Swainson's group Heleona, but is considered by that author to belong to the tribe of Sphingides, and family of Zygæidæ (Anthoceridæ, Swainson).

---

* See the Memoirs of Dr. Roxburgh in the Linnæan Transactions, and of Lieut.-Col. W. H. Sykes in the Transactions of the Royal Asiatic Society.

EUSEMIA LECTRIX.

Plate 43. fig. 2.

GENUS. EUSEMIA, *Dalman.* Phalæna, *Linn, &c.*
CH. SP. E. alis incumbentibus nigris, maculis cæruleis flavis albisque, posticis rubro alboque maculatis. Expans. alar. 3 unc.
E. with the wings incumbent, black, anterior with blue, yellow, and white spots, posterior with red and white spots. Expanse of the wings 3 inches.
SYN. Phalæna (Noctua) lectrix, *Linn. Syst. Nat.* 2. p. 834. *Fabr. Ent. Syst.* 3. 1. p. 475.
Eusemia lectrix, *Dalm. Monogr. Castn.*

This is so scarce an insect, that Mr. Drury informed Donovan he had only been able to procure a single specimen in the course of thirty years collecting insects.

EREBUS MACROPS.

Plate 44. fig. 1.

FAMILY. NOCTUIDÆ.
GENUS. EREBUS, *Latreille.* Thysania, *Dalman.*
CH. SP. E. alis dentatis, fuscis, nigro-undulatis; anticis ocello magno luteo, annulo nigro cincto. Expans. alar. 5¼ unc.
E. with the wings dentated, brown with black waves, the anterior having a large luteous ocellus, surrounded by a black ring. Expanse of the wings 5¼ inch.
SYN. Phalæna (Attacus) Macrops, *Linn Syst. Nat.* 4. p. 225.
Noctua Bubo, *Fabr. Mant. Ins.* 2. 209. *Ent. Syst.* III. 1. p. 9. Donovan, 1st Edit. (Phalæna B.) *Sulzer. Ins.* t. 22. f. 2. *Cramer. Pap.* t. 171. f. B.

This is the largest of the Chinese *Noctuæ;* some very similar species, but without the orange eye, and of a smaller size, are peculiar to China.

1. Erebus Macrops. 2. Hipparchus zonarius.
3. Callimorpha? Panthorea.

LEPIDOPTERA.

HIPPARCHUS ZONARIUS.

Plate 44. fig. 2.

FAMILY. GEOMETRIDÆ, *Leach.*
GENUS. HIPPARCHUS, *Leach.*
CH. SP. H. alis viridibus, margine posteriore late rufescente, singulis maculâ marginali viridi. Expans. alar. 1½ unc.
H. with the wings green, deeply bordered with pale red, with a green spot on the exterior margin of each wing. Expanse of the wings 1½ inch.
SYN. Phalæna Zonaria, *Donov. 1st edit.*

CALLIMORPHA? PANTHOREA.

Plate 44. fig. 3.

GENUS. CALLIMORPHA?
CH. SP. C. alis cæruleo-nigris, fasciâ maculari apicis albâ. Expans. alar. 2½ unc.
C. with blue-black wings, having a row of white spots along the posterior margins. Expanse of the wings 2¼ inches.
SYN. Phalæna Panthorea, *Cram. Ins. t.* 322. *f.* C.
Phalæna pagaria, *Fab. Ent. Syst.* III. 2. *p.* 153. *Donov. 1st edit.* (Phalæna. Geometra, p.)

The insect here figured, and those represented in plate 41. fig. 2. and plate 43. fig. 1. are very intimately allied together; nevertheless, Donovan separated them, placing one in each of the three great divisions, Sphinx, Bombyx, and Geometra.

Order. NEUROPTERA. *Linnæus.*

ÆSHNA CLAVATA.

Plate 45. fig. 1.

Family.	Libellulidæ, *Leach.*
Genus.	Æshna, *Fabr.* Libellula p. *Linn. Donov.* Cordulegaster, *Leach.*
Ch. Sp.	Æ. abdomine clavato, basi gibbo; corpore nigro, fusco viridique variegato. Expans. alar. 3½ unc.
	Æ. with the abdomen clavate gibbose at the base; body black, varied with brown and green; stigma brown. Expanse of the wings 3½ inches.
Syn.	Æshna clavata, *Fabr. Ent. Syst.* II. p. 385. *Spec. Ins.* 1. p. 526, 4.

Linnæus divides the dragon flies (Libellula, Linn) into two sections:—" 1. alis patentibus acquiescentes ;" and " 2. (alis erectis) oculi distantes remotique." Fabricius divides the Linnæan Libellulæ into three distinct genera; the first retains the Linnæan name, the second and third are called Aeshna and Agrion. Their most essential characters are taken from the form and situation of the mouth, and therefore require a deep magnifier to determine them with accuracy. Donovan states, that he had examined those parts in the greater number of the species Fabricius has described, and found his characters agree, except in one instance; which Donovan nevertheless considered a proof of the impracticability of adopting the whole of his system: he describes *Libellula Chinensis*, and refers to the only figure that has been given of it, in one of the plates of Edwards's Natural History of Birds, 1745.* Had Fabricius ever seen and examined this rare species, he must have referred it to his genus *Agrion*, each of the lips being bifid, or two-cleft, as in Libellula virgo and puella,—the essential characteristic of the genus Agrion; for the mouths of the Libellulæ of Fabricius differ altogether in structure, and are not notched in the slightest degree, as Libellula clavata, ferruginea, 6-maculata, and the European species, Libellula depressa, will sufficiently illustrate.

Donovan, however, rejecting the Fabrician generic distribution, states, that Æshna clavata must be arranged with L. grandis and forcipata; but it is nearer allied to Cordulegaster annulatus, Leach (Libellula Boltoni of Donovan's British Insects.)

---

* That Fabricius should have erred in the location of a species which he had never seen, but knew only through a rude figure, is not surprizing; but surely such a circumstance can be no proof of the impropriety of a system founded, as Donovan clearly shews, on characters of stability. J. O. W.

1. Aeshna clavata.   2. Libellula variegata.
3. Libellula 6 maculata.

NEUROPTERA.

LIBELLULA VARIEGATA.

Plate 45. fig. 2.

GENUS. LIBELLULA, *Linn. &c.*

CH. SP. L. alis fusco-maculatis et undulatis, basi flavis, posticis versus apicem maculâ magnâ fuscâ puncto flavo; apice albo. Expans. alar. 3 unc.

L. with the wings spotted and undulated with brown, yellow at the base, the posterior towards the tips with a large brown fascia, having a small yellow spot, the tips yellow. Expanse of the wings 3 inches.

SYN. Libellula variegata, *Linn. Am. Acad.* 6. 412. 86. *Syst. Nat.* 1. 2. p. 904. *Drury Exot. Ent.* 2nd edit. 2. p. 94. nec *Fabr. op. cit.* p. 382.
Libellula Histrio, *Fabr. Mant. Ins.* 1. 337. 24.
Libellula Indica, *Fabr. Ent. Syst.* 2. 376. *Guérin Icon. R. An. Ins. pl.* 60. f. 1. Donovan, 1st edit.
Libellula Arria, *Drury Exot. Entom.* 1st edit. v. 1. pl. 46. f. 1.

Another species of Libellulidæ peculiar to India, and found in China, greatly resembles this insect; it is probably a variety of it.

LIBELLULA 6-MACULATA.

Plate 45. fig. 3. ♂ et ♀

CH. SP. L. alis anticis maculis tribus costalibus atris; ultima stigmate niveo; posticis fasciis flavescentibus. Expans. alar. 1½ unc.

L. with three black spots on the costa of the anterior wings, stigma white, posterior wings with yellow clouds. Expanse of the wings 1½ inch.

SYN. Libellula 6-maculata, *Fab. Ent. Syst.* 2. p. 381.

These delicate insects appear to be male and female; they are almost a miniature resemblance of the two sexes of Libellula depressa found in Europe; one having the abdomen yellow, and the other blue.

AGRION CHINENSIS.

Plate 46. fig. 1.

GENUS. AGRION, *Fabr.* Calepteryx, *Leach.*

CH. SP. A. alis anticis testaceo-obsoletis, posticis viridibus, apice fuscis. Expans. alar. $2\frac{3}{4}$ unc.
A. with the anterior wings brownish, posterior green with brown tips. Expanse of the wings $2\frac{3}{4}$ inches.

SYN. Libellula Chinensis, *Linn. Syst. Nat.* 2. 904. 15. *Fabr. Ent. Syst.* 2. 379. *Edw. Aves t.* 112. *Guérin Icon. R. An. Ins. t.* 60. *f.* 4. (Agrion c.)

The only two specimens of this species with which Donovan was acquainted, were in the collection of the late Duchess of Portland; one of which afterwards passed into the possession of Mr. Francillon.

LIBELLULA SERVILIA.

Plate 46. fig. 2.

CH. SP. L. alis hyalinis, basi flavis; corpore rubro. Expans. alar. $2\frac{3}{4}$ unc.
L. with the wings hyaline, yellow at the base, body red. Expanse of the wings $2\frac{3}{4}$ inches.

SYN. Libellula Servilia, *Drury Exot. Ent.* 1*st edit. app. vol.* 2.
Libellula ferrugata, *Fabr. Mant. Ins.* 336. 11.
Libellula ferruginea, *Fabr. Ent. Syst.* 2. *p.* 380. *Donovan,* 1*st edit.*

Very common in China.

LIBELLULA FULVIA.

Plate 46. fig. 3.

CH. SP. L. luteo-testacea, alis fulvescentibus, marginibus anticis testaceis maculâ mediâ subpellucidâ, stigmate ad apicem fusco. Expans. alar. $2\frac{1}{3}$ unc.
L. luteo-testaceous; wings fulvescent, with the anterior margins testaceous, having a pellucid spot in the middle, and the stigma brown. Expanse of the wings $2\frac{1}{3}$ inches.

SYN. Libellula Fulvia, *Drury Exot. Ent. vol.* 2. *pl.* 46. *fig.* 2.

1. Agrion Chinensis.   2. Libellula Serivlia!
3. Libellula Fulvia.

Pl. 47

Epeira Maculata.

Order. **DIMEROSOMATA**. *Leach.*

EPEIRA (NEPHILA) MACULATA.
Plate 47.

CLASS.	ARACHNIDA. *Lamarck.* Aptera p. *Linnæus.*
ORDER.	DIMEROSOMATA, *Leach.*
FAMILY.	ARANEIDÆ, *Leach.*
GENUS.	EPEIRA, *Walckenäer.*
SUB-GEN.	NEPHILA, *Leach Zool. Misc.*
CH. SP.	E. corpore elongato, cephalo-thorace holosericeo argenteo, abdomine cylindrico fusco-rubro lineis punctisque albis; pedibus longissimis atris. Long. Corp. 1¾ unc.
	E. with the body elongated, cephalo-thorax holosericeus and silvery, abdomen cylindric, red-brown with spots and lines of white, legs very long and black. Length of the body 1¾ inches.
SYN.	Aranea maculata, *Fabr. Ent. Syst.* 2. p. 425.

This remarkable creature is peculiar to some parts of the Chinese empire. It is not the largest of the genus known; yet it is of sufficient magnitude to excite terror and disgust. To an European, who has seen only the indigenous spiders of his own country, a species five or six inches in length, and nearly the same in breadth, must appear a frightful creature: Epeira Maculata sometimes exceeds that size; but it has not the forbidding aspect of most insects of the same genus. The legs are unusually long, and the body slender. In its general appearance it resembles some kinds of the *Phalangia* that are known in England by the vulgar name *Harvest-men*, being generally seen about that time of the year.

It has been observed, that nature oftentimes adorns the most deformed and loathsome of her creatures in the richest display of colours; and this is especially noticed in many sorts of snakes, toads, lizards, &c. Spiders seem also of this description: to a form the most hideous we frequently find united a brilliance of colours, and elegance of marking, that is scarcely excelled by any of the butterfly tribe,—the most beautiful of all lepidopterous insects. Our present subject is a striking proof of the latter part of this observation. The three figures in our plate of Epeira Maculata exhibit a front and a profile view of the insect, together with the front of the head at the third figure. The head is furnished with two very strong black mandibles, each terminated in an extremely acute point. The fore part of the cephalo-thorax, which is wholly of a fine silky appearance, and the colour of silver, bending over the mandibles in the form of an arch, or circular

DECAPODA.

head-piece, gives it the resemblance of a black head with a crown of silver on the brow. This appearance is heightened in no small degree by three rugged prominences, one in the centre, and another on each side, on the upper part; and by the minute black eyes, which, like those of most spiders, sparkle with the lustre of small gems. These eyes are eight in number, four are placed immediately in the front of the silver-coloured circular front piece, and on each side are two placed close together thus : ∵ :

The body is really beautiful, the chief colour is deep brown, strongly tinged with bright purple; a broad stripe of orange colour passes down the abdomen from the cephalo-thorax to the apex : the whole is elegantly marked with a variety of cream-coloured lines and spots intersecting each other. Very little hair is found on any part of this spider except on the cephalo-thorax, which being rubbed off, discovers a hard testaceous black substance beneath.

The description given by Fabricius accords in every respect with our specimen. The only insect with which it could possibly be confounded is Aranea Pilipes, which also has never been figured; it differs, however, from Aranea Maculata in the very hairy clothing of the legs, and it has also two silver stripes down the back : a striking specific distinction to separate it from our insect. It is also a native of the East Indies, but not of China, that we are informed.

Order. DECAPODA. *Latreille.*

ORITHYIA MAMILLARIS.

Plate 48.

CLASS.	CRUSTACEA, *Cuvier.* Aptera p. *Linnæus.*
ORDER.	DECAPODA, *Latreille.*
SECTION.	BRACHYURA, *Latreille.*
FAMILY.	PORTUNIDÆ.
GENUS.	ORITHYIA. *Fabricius.* Cancer p. *Linn. Donov.*
CH. SP.	O. testâ utrinque trispinosâ, taberculatâ, maculis duabus rufis, fronte tridentato. Long. test. 1¼ unc.
	O. with the carapax having three spines on each side and tuberculated, with two red spots, the front tridentate. Length 1¼ inch.
SYN.	Cancer mamillaris, *Fabr. Ent. Syst.* 2. 465. 91.

It is worthy of remark, that this is the only species of the old genus *Cancer*, which Fabricius mentions as a native of China.

Orithyia mamillaris.

Pl. 49.

Squilla Mantis.

Order. STOMAPODA. *Latreille.*

SQUILLA MANTIS.

Plate 49.

GENUS. SQUILLA, *Fabricius.* Cancer, p. *Linnæus.*

CH. SP. Sq. manibus compressis 6-dentatis, articulo ultimo abdominis carinâ centrali, dentibus tribus utrinque apiceque bidentato. Long. Corp. $4\frac{1}{2}$ unc.

Sq. with the terminal joint of the large pair of claws six toothed, the last abdominal segment with a central carina, three teeth on each side and two at the extremity. Length of the body $4\frac{1}{2}$ inches.

SYN. Cancer Mantis, *Linn. Syst. Nat.* 2. 1054. *Fab. Ent. Syst.* 2. 511. *Desmarest Cons. sur les Crustaces, p.* 251. *Encycl. Méth. pl.* 324. *De Geer Ins. vol.* 7. *t.* 34.

The Linnæan *Cancri* are numerous, and include many species not less singular in appearance than the extraordinary creature before us. Indeed, some species are so extremely different from the rest, both in structure and manners of life, that even Donovan could not hesitate in concluding the Linnæan character of the genus to be defective and indefinite. This may be observed in several of the species Linnæus himself described, and throughout a more extensive number of those discovered since the time of that author. It is evident Linnæus could never reconcile the subdivisions of the two principal families, the *Brachyuri* and *Macrouri*, or crabs with short and long tails; and later naturalists have ventured, with propriety, to alter this part of his arrangement.

Desmarest gives the Mediterranean as the locality of Squilla Mantis, but the species here figured, which is very common in the Chinese boxes of insects sent to this country, agrees with the extended characters given of S. Mantis. Fabricius says of it, " Habitat in mari *Asiatico, Indico, Mediterraneo,* Italis esculentus."

MYRIAPODA.

SCOLOPENDRA MORSITANS?

Plate 50.

CLASS.	AMETABOLA, *Leach.* Myriapoda, *Latreille.* Aptera p. *Linnæus.*
FAMILY.	SCOLOPENDRIDÆ.
GENUS.	SCOLOPENDRA, *Linnæus.*
CH. SP.	Sc. pedibus utrinque 21, posterioribus spinosis. Long. Corp. 6 unc. Sc. with twenty-one feet on each side, the posterior pair spined. Length of the body 6 inches.
SYN.	Scolopendra morsitans, *De Geer Ins. vol.* 7. *t.* 43. *fig.* 1. (in *Indiâ*). *Linn.? Syst. Nat.* 2. 1063. ("Hab. in *Indiis*"). *Fabricius Ent. Syst.* 2. p. 390. ("Hab. in *Indiâ orientali.*")

Travellers agree that the temperate parts of Asia would be a terrestrial paradise, were it not for the multitude of troublesome insects and reptiles with which they are infested. In a well cultivated country like China, many of these creatures can scarcely find shelter; but such as harbour in the walls or furniture of human dwellings are as abundant in that, as any other country lying within or near the tropics. Amongst the latter, none produce more terrible effects than the Centipede, whose poison is as venomous as that of the scorpion, which is also a native of China.

Sir G. Staunton mentions a remarkable circumstance that occurred during the embassy to China to which he was attached. The ambassador and his suite were accommodated in a temple near the suburbs of Tong-choo-foo. "In some of the apartments the priests had suffered scorpions and scolopendras to harbour through neglect. These noisome creatures were known only by description to some of the gentlemen in the embassy, who had not visited the southern parts of Europe: the sight of such, for the first time, excited a degree of horror in their minds; and it seemed to them to be a sufficient objection to the country, that it produced these animals." Sir George however adds, that no accident happened in that instance.—The species of Scolopendra he alludes to, is probably *Morsitans*, which is common in many parts of the world, but is particularly found of a frightful size, and in vast abundance, in the two Indies.

Many authors have described this creature. In the voluminous works of Seba we find several specimens of it from different countries, differing materially in size, and some trifling particulars. The largest of these exceed our figure in magnitude, being near fourteen inches in length: this he calls *Millepeda major* ex nova Hispania. His figure of *Millepeda Africana* is about the size of our Chinese specimen. He has also a third

Scolopendra Morsitans.

and fourth figure, *Millepeda Orientalis* and *Millepeda Ceylonica, mas:* the latter is the same length as our figure, but the body is very narrow. Millepeda Orientalis is also the same length, but the body is very broad. Some of these insects are not four inches in length. These will be regarded as distinct species; and, indeed, it is questionable whether the Chinese species be strictly entitled to the specific name of Morsitans, as Guérin (Encyclop. Méth. x. p. 393), and Pohl and Kollar, in their work on the noxious insects of Brazil, have given the name of Morsitans to the Brazilian species, which has also twenty-one pairs of legs; to this species, however, Dr. Leach gave the specific name of Alternans. The entire genus has, indeed, need of a monographical revision. It will be seen that the antennæ in this figure are much shorter than in that of De Geer, &c.

Authors agree that they vary exceedingly in size* and colour. De Geer describes them to be sometimes deep reddish brown; at others, the colour of yellow ochre. The figure in Catesby's Natural History of Carolina is light brown; we have specimens of a livid yellow, and have seen others strongly tinged with red.

The last pair of legs is considerably larger than the others, and is armed with small black spines. The legs terminate in very sharp hooks or nails of a shining black colour. All the other legs are also furnished with a smaller nail of the same shape and colour.

M. Gronovius says, that all its feet are very venomous; but the most formidable of its weapons are the two sharp hooked instruments that are placed under the mouth, and with which it destroys its prey.

Leuwenhoeck having examined these instruments with a microscope, found a small opening at the extremity of each, and a channel from them into the body of the creature. Through this channel he supposes the Scolopendra emits the poisonous fluid into the wound it makes with the hooked instrument. That author further remarks, that he has seen a liquor on that part of living scolopendras. A figure of these instruments on the under side of the head is represented in one of the dissections in our plate.

The same author, wishing to ascertain the influence of the poison of Scolopendra morsitans, placed a large fly within its reach. The Scolopendra at first took it between a pair of its middle feet, then passed it from one pair of feet to the next, till the fly was brought under the sharp pointed instrument or crotchets at the head, which it plunged into the fly, and it died instantly. Notwithstanding this experiment, De Geer, Catesby, and other authors assert, that its bite seldom proves fatal to larger animals; but all agree

---

* These creatures differ from most insects in their manner of growth, insomuch that it is impossible to ascertain when they are of their full size. The segments of the body increase in number as they advance in age, which circumstance renders it difficult oftentimes to determine the species without a minute examination of its other characters.

MYRIAPODA.

that its poison is as dangerous as that of the scorpion. (See also Worbe in the Bulletin de la Soc. Philomat, Jan. 1824, Amoreux Insectes venimeux, and the recent work of Pohl and Kollar above referred to, for further details relative to the poisonous properties of these insects.)

This Scolopendra has eight eyes: they are very small; four are placed on each side of the head near the antennæ. In the dissections a figure is given to exhibit the manner in which the four eyes are placed on one side.

ALPHABETICAL INDEX.

In this Index the names employed both in the present and former editions are introduced, in order to render the references which have been made by writers to the former edition available. The names, generic, sub-generic, or specific, first employed in the present edition are distinguished by a *.

*Acræa Vesta, Pl. 30. f. 1.
*Agrion Chinensis, Pl. 46. f. 1.
Aranea maculata, Pl. 47.
*Argynnis Erymanthis, Pl. 35. f. 1.
*Astemma Schlanbuschii, Pl. 20. f. 2.
*Æshna clavata, Pl. 45. f. 1.
*Belostoma *Indica ? Pl. 18.
*Belostoma *(Sphærodema) rustica, Pl. 19. f. 1.
Buprestis ocellata, Pl. 7. f. 2.
Buprestis *(Chrysochroa) ocellata, Pl. 7. f. 2.
Buprestis vittata, Pl. 7. f. 1.
Buprestis *(Chrysochroa) vittata, Pl. 7. f. 1.
*Calandra longipes, Pl. 4. f. 2.
*Callidea *ocellata, Pl. 20. f. 1.
*Callidea Stockerus, Pl. 21. f. 1.
*Callimorpha? bifasciata, Pl. 41. f. 4.
*Callimorpha? *panthorea, Pl. 44. f. 3.
*Callimorpha? ruficollis, Pl. 41. f. 3.
*Callimorpha? Thallo, Pl. 41. f. 2.
Cancer Mamillaris, Pl. 48.
Cancer Mantis, Pl. 49.
Cerambyx farinosus, Pl. 6. f. 3.
Cerambyx reticulator, Pl. 6. f. 2.
Cerambyx Rubus, Pl. 6. f. 1.
*Cerbus *tenebrosus? Pl. 21. f. 4.
*Cercopis abdominalis, Pl. 16. f. 5.
Cetonia Chinensis, Pl. 3. f. 1.
Cetonia *(Tetragona) Chinensis, Pl. 3. f. 1.
*Chariesterus cruciger, Pl. 21. f. 3.
Cicada abdominalis, Pl. 16. f. 5.
Cicada ambigua, Pl. 16. f. 2.
Cicada atrata, Pl. 15.
Cicada frontalis, Pl. 16. f. 6.

Cicada lanata, Pl. 16. f. 3.
Cicada limbata var. Pl. 17.
Cicada sanguinea, Pl. 16. f. 1.
Cimex aurantius, Pl. 21. f. 2.
Cimex bifidus, Pl. 21. f. 5.
Cimex cruciger, Pl. 21. f. 3.
Cimex dispar, Pl. 20. f. 1.
Cimex Phasianus, Pl. 21. f. 4.
Cimex Slanbuschii, Pl. 20. f. 2.
Cimex Stockerus, Pl. 21. f. 1.
*Cleonis perlatus, Pl. 4. f. 7.
*Colias *(Callidryas) Philea, Pl. 32. f. 2.
*Colias *(Callidryas) Pyranthe, Pl. 32. f. 1.
*Copris Bucephalus, Pl. 2. f. 3.
*Copris Midas, Pl. 1. f. 1.
*Copris Molossus, Pl. 2. f. 1.
Curculio barbirostris, Pl. 4. f. 3.
Curculio Chinensis, Pl. 4. f. 1.
Curculio longipes, Pl. 4. f. 2.
Curculio perlatus, Pl. 4. f. 7.
Curculio pulverulentus, Pl. 4. f. 6.
Curculio squamosus, Pl. 4. f. 5. and Pl. 5.
Curculio verrucosus, Pl. 4. f. 4.
*Cynthia Almana, Pl. 36. f. 2.
*Cynthia Œnone, Pl. 36. f. 1.
*Cynthia Orithya, Pl. 35. f. 2.
*Deilephila Nechus, Pl. 40. f. 1.
*Epeira *(Nephila) maculata, Pl. 47.
*Erebus *macrops, Pl. 44. f. 1.
*Euchlora viridis, Pl. 3. f. 2.
*Eusemia lectrix, Pl. 43. f. 2.
*Flata *nigricornis, Pl. 17.
Fulgora Candelaria, Pl. 14.

ALPHABETICAL INDEX.

*Glaucopis Polymena, Pl. 40. f. 2.
*Gryl otalpa *Chinensis, Pl. 12. f. 2.
Gryllus acuminatus, Pl. 11. f. 2.
Gryllus *(Conocephalus) acuminatus? Pl. 11. f. 2.
Gryllus flavicornis, Pl. 12. f. 1.
Gryllus Gryllotalpa, Pl. 12. f. 2.
Gryllus morbillosus, Pl. 13.
Gryllus nasutus, Pl. 10. f. 1.
Gryllus perspicillatus, Pl. 11. f. 1.
Gryllus *(Phasgoneurus) perspicillatus, Pl. 11. f. 1.
Gryllus vittatus, Pl. 10. f. 2.
*Gymnopleurus *sinuatus, Pl. 1. f. 5.
*Harpactor bifidus, Pl. 21. f. 5.
*Heleona militaris, Pl. 43. f. 1.
*Hipparchus Zonarius, Pl. 44. f. 2.
*Hipporhinus verrucosus, Pl. 4. f. 4.
*Hypomeces squamosus, Pl. 4. f. 5. and Pl. 5.
*Hypomeces squamosus *var. Pl. 4. f. 6.
*Lamia *punctator, Pl. 6. f. 3.
*Lamia reticulator, Pl. 6. f. 2.
*Lamia Rubus, Pl. 6. f. 1.
Libellula Chinensis, Pl. 46. f. 1.
Libellula clavata, Pl. 45. f. 1.
Libellula ferruginea, Pl. 46. f. 2.
Libellula Fulvia, Pl. 46. f. 3.
Libellula Indica, Pl. 45. f. 2.
Libellula *Servilia, Pl. 46. f. 2.
Libellula 6-maculata, Pl. 45. f. 3.
Libellula *variegata, Pl. 45. f. 2.
*Limenitis *Eurynome, Pl. 35. f. 4.
*Limenitis *Leucothoe, Pl. 35. f. 3.
*Locusta *(Rutidoderes) flavicornis, Pl. 12. f. 1.
*Locusta *(Phymatea) morbillosa, Pl. 13.
*Lystra lanata, Pl. 16. f. 3.
Mantis *(Schizocephala) *bicornis, Pl. 9. f. 1.
Mantis flabellicornis, Pl. 9. f 2.
Mantis *(Empusa) flabellicornis, Pl. 9. f. 2.
Mantis oculata, Pl. 9. f. 1.
Meloe Cichorei, Pl. 8. f. 1.
Melolontha viridis, Pl. 3. f. 2.
*Morpho *(Drusilla) Jairus, Pl. 33.
*Morpho Rhetenor, Pl. 29.
*Mylabris Cichorii, Pl. 8. f. 1.
*Myrina *(Loxura) Atymnus, Pl. 39. f. 1.
Nepa grandis, Pl. 18.
Nepa rubra, Pl. 19. f. 2.
Nepa rustica, Pl. 19. f. 1.

*Nymphalis Antiochus, Pl. 37. f. 2.
*Nymphalis *(Charaxes) Bernardus, Pl. 34.
*Nymphalis Jacintha, Pl. 37. f. 1.
*Nymphalis *(Aconthea) Lubentina, Pl. 36. f. 3.
*Nymphalis *Sylla, Pl. 38.
*Oniticellus cinctus, Pl. 1. f. 3.
*Onthophagus seniculus, Pl. 2. f. 2.
*Orithyia mamillaris, Pl. 48.
*Oryctes *Rhinoceros? Pl. 1. f. 2.
Papilio Agamemnon, Pl. 26. f. 2.
Papilio Agenor, Pl. 24. f. 2.
Papilio Almana, Pl. 36. f. 2.
Papilio Antiochus, Pl. 37. f. 2.
Papilio Atymnus, Pl. 39. f. 1.
Papilio Bernardus, Pl. 34.
Papilio Coon, Pl. 24. f. 1.
Papilio Crino, Pl. 23.
Papilio Demoleus, Pl. 28. f. 2.
Papilio Epius, Pl. 28. f. 1.
Papilio Erymanthis, Pl. 35. f. 1.
Papilio Gambrisius, Pl. 38.
Papilio Glaucippe, Pl. 31. f. 1.
Papilio Hyparete, Pl. 30. f. 3.
Papilio Jacintha, Pl. 37. f. 1.
Papilio Jairus, Pl. 33.
Papilio Laomedon, Pl. 27.
Papilio Leucothoe, Pl. 35. f. 4.
Papilio Lubentina, Pl. 36. f. 3.
Papilio Mæcenas, Pl. 39. f. 2.
Papilio Œnone, Pl. 36. f. 1.
Papilio Orythia, Pl. 35. f. 2.
Papilio Paris, Pl. 22.
Papilio Pasithoe, Pl. 30. f. 2.
Papilio Peranthus, Pl. 25.
Papilio Philea, Pl. 32. f. 2.
Papilio Polyxena, Pl. 35. f. 3.
Papilio *Protenor, Pl. 27.
Papilio Pyranthe, Pl. 32. f. 1.
Papilio Rhetenor, Pl. 29.
Papilio Sesia, Pl. 31. f. 2.
Papilio Telamon, Pl. 26. f. 1.
Papilio Vesta, Pl. 30. f. 1.
Phalæna Atlas, Pl. 42.
Phalæna bubo, Pl. 44. f. 1.
Phalæna lectrix, Pl. 43. f. 2.
Phalæna militaris, Pl. 43. f. 1.
Phalæna pagaria, Pl. 44. f. 3.

ALPHABETICAL INDEX.

Phalæna zonaria, Pl. 44. f. 2.
*Pieris *(Iphias) Glaucippe, Pl. 31. f. 1.
*Pieris Hyparete, Pl. 30. f. 3.
*Pieris Pasithoe, Pl. 30. f. 2.
*Pieris *(Thestias) *Pyrene? Pl. 31. f. 2.
*Raphigaster aurantius, Pl. 21. f. 2.
*Rhina barbirostris, Pl. 4. f. 3.
*Rhinastus *sternicornis, Pl. 4. f. 1.
*Sagra *splendida, Pl. 8. f. 2.
*Saturnia Atlas, Pl. 42.
Scarabæus Bucephalus, Pl. 2. f. 3.
Scarabæus cinctus, Pl. 1. f. 3.
Scarabæus Leei, Pl. 1. f. 5.
Scarabæus Midas, Pl. 1. f. 1.
Scarabæus Molossus, Pl. 2. f. 1.
Scarabæus nasicornis, Pl. 1. f. 2.
Scarabæus sacer, Pl. 1. f. 4.

Scarabæus *(Heliocantharus) *sanctus? Pl. 1. f. 4.
Scarabæus seniculus, Pl. 2. f. 2.
Scolopendra morsitans, Pl. 50.
*Sesia Hylas, Pl. 41. f. 1.
Sphinx bifasciata, Pl. 41. f. 4.
Sphinx Hylas, Pl. 41. f. 1.
Sphinx Nechus, Pl. 40. f. 1.
Sphinx Polymena, Pl. 40. f. 2.
Sphinx ruficollis, Pl. 41. f. 3.
Sphinx Thallo, Pl. 41. f. 2.
*Squilla mantis, Pl. 49.
Tenebrio femoratus, Pl. 8. f. 2.
*Tettigonia frontalis, Pl.16. f. 6.
Tettigonia splendidula, Pl. 16. f. 4.
*Thecla Mæcenas, Pl. 39. f. 2.
*Truxalis *Chinensis, Pl. 10. f. 1.
*Truxalis *(Mesops) vittatus, Pl. 10. f. 2.

SYSTEMATIC INDEX.

INSECTA.

I.—MOUTH WITH JAWS.

Order. COLEOPTERA.
Tribe. LAMELLICORNES, *Latr.*
Family. SCARABÆIDÆ, *Mac L.*

Scarabæus (Heliocantharus) sanctus? Pl. 1. f. 4.
Gymnopleurus sinuatus, Pl. 1. f. 5.
Copris Midas, Pl. 1. f. 1.
C.   Molossus, Pl. 2. f. 1.
C.   Bucephalus, Pl. 2. f. 3.
Onthophagus seniculus, Pl. 2. f. 2.
Oniticellus cinctus, Pl. 1. f. 3.

Family. DYNASTIDÆ, *Mac L.*

Oryctes Rhinoceros, Pl. 1. f. 2.

Family. MELOLONTHIDÆ, *Mac L.*

Euchlora viridis, Pl. 3. f. 2.

Family. CETONIIDÆ, *Mac L.*

Cetonia (Tetragona) Chinensis, Pl. 3. f. 1.

Tribe. STERNOXI, *Latr.*
Family. BUPRESTIDÆ, *Leach.*

Buprestis (Chrysochroa) vittata, Pl. 7. f. 1.
——————————— ocellata, Pl. 7. f. 2.

Tribe. TRACHELIDES, *Latr.*
Family. MELOIDÆ, *Westw.*

Mylabris Cichorii, Pl. 8. f. 1.

Tribe. RHYNCOPHORA, *Latr.*
Family. CURCULIONIDÆ, *Leach.*

Hipporhinus verrucosus, Pl. 4. f. 4.
Hypomeces squamosus, Pl. 4. f. 5. and Pl. 4. f. 6.
——————————— var. Pl. 5.

Cleonis perlatus, Pl. 4. f. 7.
Rhinastus sternicornis, Pl. 4. f. 1.
Rhina barbirostris, Pl. 4. f. 3.
Calandra longipes, Pl. 4. f. 2.

Tribe. LONGICORNES, *Latr.*
Family. LAMIIDÆ, *Westw.*

Lamia Rubus, Pl. 6. f. 1.
L.   reticulator, Pl. 6. f. 2.
L.   punctator, Pl. 6. f. 3.

Tribe. EMPODA.
Family. SAGRIDÆ, *Westw.*

Sagra splendida, Pl. 8. f. 2.

Order. ORTHOPTERA. *Oliv.*
Tribe. CURSORIA, *Latr.*
Family. MANTIDÆ, *Leach.*

Mantis (Empusa) flabellicornis, Pl. 9. f. 2.
M.   (Schizocephala) bicornis, Pl. 9. f. 1.

Tribe. SALTATORIA, *Latr.*
Family. LOCUSTIDÆ, *Leach.*

Truxalis Chinensis, Pl. 10. f. 1.
T.   (Mesops) vittatus, Pl. 10. f. 2.
Locusta (Rutidoderes) flavicornis, Pl. 12. f. 1.
L.   (Phymatea) morbillosa, Pl. 13.

Family. GRYLLIDÆ, *Leach.*

Gryllus (Phasgonurus) perspicillatus, Pl. 11. f. 1.
G.   (Conocephalus) acuminatus? Pl. 11. f. 2.

Family. ACHETIDÆ, *Leach.*

Gryllotalpa Chinensis, Pl. 12. f. 2.

SYSTEMATIC INDEX.

Order. NEUROPTERA. *Linn.*
Family. LIBELLULIDÆ, *Leach.*
Æshna clavata, Pl. 45. f. 1.
Libellula variegata, Pl. 45. f. 2.
Libellula 6-maculata, Pl. 45. f. 3.
L.   Servilia, Pl. 46. f. 2.
L.   Fulvia, Pl. 46. f. 3.
Agrion Chinensis, Pl. 46. f. 1.

II.—MOUTH SUCTORIAL.

Order. LEPIDOPTERA. *Linn.*

Tribe. DIURNA, *Latr.*
Family. PAPILIONIDÆ, *Leach.*

Papilio Paris, Pl. 22.
P.   Crino, Pl. 23.
P.   Coon, Pl. 24. f. 1.
P.   Agenor, Pl. 24. f. 2.
P.   Peranthus, Pl. 25.
P.   Telamon, Pl. 26. f. 1.
P.   Agamemnon, Pl. 26. f. 2.
P.   Protenor ♀, Pl. 27.
P.   Epius, Pl. 28. f. 1.
P.   Demoleus, Pl. 28. f. 2.
Pieris Pasithoe, Pl. 30. f. 2.
P.   Hyparete, Pl. 30. f. 3.
P.   (Iphias) Glaucippe, Pl. 31. f. 1.
P.   (Thestias) Pyrene? Pl. 31. f. 2.
Colias (Callidryas) Pyranthe, Pl. 32. f. 1.
C.   (C.   ) Philea, Pl. 32. f. 2.

Family. HELICONIIDÆ, *Swainson.*
Acræa Vesta, Pl. 30. f. 1.

Family. NYMPHALIDÆ, *Swainson.*
Morpho Rhetenor, Pl. 29.
M.   (Drusilla) Jairus, Pl. 33.
Nymphalis Jacintha, Pl. 37. f. 1,
N.   Antiochus, Pl. 37. f. 2.
N.   Sylla, Pl. 38.
N.   (Charaxes) Bernardus, Pl. 34.
N.   (Aconthea) Lubentina, Pl. 36. f. 3.
Argynnis Erymanthis, Pl. 35. f. 1.
Cynthia Orithya, Pl. 35. f. 2.
C.   Œnone, Pl. 36. f. 1.
C.   Almana, Pl. 36. f. 2.
Limenitis Leucothoe, Pl. 35. f. 3.
L.   Eurynome, Pl. 35. f. 4.

Family. LYCÆNIDÆ, *Swainson.*
Myrina (Loxura) Atymnus, Pl. 39. f. 1.
Thecla Mæcenas, Pl. 39. f. 2.

Section. CREPUSCULARIA, *Latr.*
Family. SPHINGIDÆ, *Leach.*
Deilephila Nechus? Pl. 40. f. 1.
Sesia Hylas, Pl. 41. f. 1.

Section. ———?
Family. ZYÆNIDÆ, *Leach.*
Glaucopis Polymena, Pl. 40. f. 2.

Section. NOCTURNA, *Latr.*
Family. BOMBYCIDÆ, *Leach.*
Saturnia Atlas, Pl. 42.
Eusemia Lectrix, Pl. 43. f. 2.

Family. ARCTIIDÆ, *Stephens.*
Callimorpha? Thallo, Pl. 41. f. 2.
C.   ? ruficollis, Pl. 41. f. 3.
C.   ? bifasciata, Pl. 41. f. 4.
C.   ? panthorea, Pl. 44. f. 3.
Heleona militaris, Pl. 43. f. 1.

Family. NOCTUIDÆ, *Leach.*
Erebus macrops, Pl. 44. f. 1.

Family. GEOMETRIDÆ, *Leach.*
Hipparchus Zonarius, Pl. 44. f. 2.

Order. HEMIPTERA. *Linn.*
Sub-Order. HETEROPTERA, *Latr.*
Section. GEOCORISA, *Latr.*
Family. SCUTELLERIDÆ, *Leach.*
Callidea ocellata, Pl. 20. f. 1.
C.   Stockerus, Pl. 21. f. 1.
Raphigaster aurantius, Pl. 21. f. 2.

SYSTEMATIC INDEX.

Family. COREIDÆ, *Leach.*

Chariesterus cruciger, Pl. 21. f. 3.
Cerbus tenebrosus? Pl. 21. f. 4.

Family. LYGÆIDÆ, *Leach.*

Astemma Schlanbuschii, Pl. 20. f. 2.

Family. REDUVIIDÆ, *Leach.*

Harpactor bifidus, Pl. 21. f. 5.

Section. HYDROCORISA, *Latr.*
Family. NEPIDÆ, *Leach.*

Belostoma Indica? Pl. 18.
B.   (Sphærodema) rustica, Pl. 19. f. 1.
Nepa rubra, Pl. 19. f. 2.

Sub-Order. HOMOPTERA, *Latr.*
Family. CICADIDÆ, *Leach.*

Cicada atrata, Pl. 15.
C.   sanguinea, Pl. 16. f. 1.
C.   ambigua, Pl. 16. f. 2.
C.   splendidula, Pl. 16. f. 4.

Family. FULGORIDÆ, *Leach.*

Fulgora Candelaria, Pl. 14.
Lystra lanata, Pl. 16. f. 3.
Flata nigricornis, Pl. 17.

Family. CERCOPIDÆ, *Leach.*

Cercopis abdominalis, Pl. 16. f. 5.
Tettigonia frontalis, Pl. 16. f. 6.

CRUSTACEA.

Order. DECAPODA.

Orithyia mamillaris, Pl. 48.

Order. STOMAPODA.

Squilla mantis, Pl. 49.

ARACHNIDA.

Epeira (Nephila) clavipes, Pl. 47.

AMETABOLA.

Scolopendra morsitans, Pl. 50.

THE END.

C. Whittingham, Tooks Court, Chancery Lane.

NATURAL HISTORY

OF THE

INSECTS OF INDIA,

CONTAINING

UPWARDS OF TWO HUNDRED AND TWENTY

FIGURES AND DESCRIPTIONS,

BY

E. DONOVAN, F.L.S. & W.S.

A NEW EDITION,

BROUGHT DOWN TO THE PRESENT STATE OF THE SCIENCE,
WITH SYSTEMATIC CHARACTERS OF EACH SPECIES, SYNONYMS, INDEXES, AND OTHER
ADDITIONAL MATTER,

BY J. O. WESTWOOD,

SECRETARY OF THE ENTOMOLOGICAL SOCIETY OF LONDON, HON. MEM. OF THE LITERARY AND HISTORICAL
SOCIETY OF QUEBEC, AND OF THE NATURAL HISTORY SOCIETIES OF MOSCOW, LILLE, MAURITIUS, ETC.

LONDON:

HENRY G. BOHN, 4 & 5, YORK STREET, COVENT GARDEN.

MDCCCXLII.

PREFACE.

At the period when the first edition of this work was presented to the public, the study of exotic insects, and indeed the science of Entomology itself, had made but little progress in this country. The collections of Francillon, Drury, MacLeay, Sir J. Banks, and Donovan, contained almost all that was then known of Indian Entomology, with which our Continental neighbours were then, as still, comparatively ignorant. To these collections, examined by Fabricius himself, Donovan had free access, and his figures of the insects therein contained, which had served as types for the descriptions of the Entomologist of Kiel, are especially valuable.

The progress of Entomology, as a science, has so much advanced, as to render a republication of this work advisable; at the same time, however, requiring that its original Linnæan style should not be retained, but that it should be brought down to the present state of science. This I have endeavoured to do, by rendering the specific characters more detailed, the nomenclature more correct, the synonyms more numerous, and the localities more precise. I have added many additional observations, omitting nothing which appeared in the former at all likely to instruct or interest the reader. Alphabetical and systematic indices of the work are introduced, as well as numbers, both for the plates, and for the individual figures on each plate, which were omitted in the former edition, which appeared in parts, commencing in the year 1800.

It has unfortunately happened, from the careless indications of the older authorities, that many insects inhabiting the West Indies have been given as natives of East India; and hence it has happened that Donovan, having no means of ascertaining the true locality of various species, introduced into the present work several West Indian insects. With these exceptions, the present work is intended to illustrate " the Entomological productions of a country for which we ought to cherish the liveliest and deepest interest, as being connected intimately with the prosperity, the dignity, and the honour of the British Empire; in a word, of British India" embracing also illustrations of those species which inhabit every other part of that vast continent, as well as the islands situated in the Indian Seas.

In respect, therefore, both to the circumstance of this great territory being peculiarly confided to our care, and more especially to the remarkable character of its natural productions, the investigation of its Entomological treasures becomes the especial province of the English entomologist.

Thus, whilst our museums teem with undescribed insects from India, collected by General Hardwicke, Colonel Sykes, Colonel Whithill, Captain Smee, Messrs. Saunders, Royle, Downes, &c. &c., Dr. Perty, writing in 1831, observes in his " Observationes nonnullæ in Coleoptera Indiæ Orientalis,"—" Præsertim tempore novissimo in Galliæ et Germaniæ museis, communicatione propinqua cum illa regione deficiente, Indiæ orientalis Coleoptera rarius inveniuntur."

In order to shew the peculiar character of Indian Entomology to its full extent, a far greater space would be required than can here be possibly given to it. A few remarks will not, however, be out of place.

From its peculiar situation, as the great intermediate southern peninsula of Asia, it may be easily conceived, not only that India comprises various types of form peculiar to itself, but that it also borrows portions of those found in the Arabian and Siamese peninsulæ; the former of which comprises, of course, a greater portion of forms peculiar to eastern Africa, Asia minor, and even southern Turkey; whilst to the latter is imparted a portion of the peculiarities of the Entomology of Borneo, and the other great islands lying south of the Equator. The entomological peculiarities of the Himalayan Mountains have been submitted to a minute analysis, by the Rev. F. W. Hope.

In the order Coleoptera we accordingly find various genera and species, either exclusively confined to India, or occurring therein and in the countries above indicated. Thus the most splendid species of Cicindela, and the genera Therates, Tricondyla, and Colliuris, in the family Cicindelidæ, do not occur beyond the limits of India and the Indian Archipelago. The genera Catascopus, Orthogonius, many fine Panagæi, &c. amongst the Carabidæ; the most splendid of all the Buprestidæ, and the beautiful group of Elateridæ, typified by Elater aureolus, (some of the species of both of which extend in their range to China,) occur in India. In the great group of Lamellicorn Coleoptera, (Scarabæus, *Linn.*) the species of Onthophagus are excessively numerous, of large size and fine colours, whilst the giant Dynastidæ, (which are so abundant in South America,) are here represented chiefly by the small group Chalcosoma, *Hope*, (S. Atlas, &c.) and by D. Dichotomus and D. longimanus; the Cetoniæ, Euchloræ, and Popilliæ, on the other hand, are far more numerous and beautiful. The Lucani are of much larger size, and far more numerous than in South America. The species of Longicorn beetles are numerous, but the giant Prioni are of very rare occurrence in India, as compared with South America, The splendid genera Sagra, Podontia, and Phyllocharis, with many fine Eumorphi, and various Paussidæ, are especially natives of these regions.

In the Orthoptera, many curious Phasmæ, with the singular genus Phyllium, and numerous splendid Grylli, *Linn.*, including the remarkable Schizodactyla monstrosa, may be mentioned. In the Hemiptera, the most splendid of all the species of Scutellera, and of Cicada, with several curious Fulgoræ, and other Fulgorideous insects;[*]

[*] M. Guérin has recently elucidated this group, in Bélanger's "Voyage aux Indes Orientales."

and in the Lepidoptera, Priamus, and its allies, (the most magnificent of all the butterflies, *Ornithopterus*,) and the delicate Ideæ, are especially to be noticed, as well as several species of the curious Dipterous genus Diopsis.

We are still, however, far from having attained a perfect idea of the Entomological treasures of India, every new arrival making us acquainted with new and beautiful species.

INDIA.

Order. COLEOPTERA. *Linnæus.*

DYNASTES (CHALCOSOMA) ATLAS.

Plate I.

TRIBE. LAMELLICORNES, *Latreille.*

FAMILY. DYNASTIDÆ, *MacLeay.*

GENUS. DYNASTES, *MacLeay.* Scarabæus, *Latreille.* Geotrupes, *Fabricius.* (SUBGENUS: Chalcosoma, *Hope.*)

SPECIES. DYNASTES ATLAS: cupreo-niger, thorace tricorni; anteriori brevissimo; capitis cornu longissimo adscendente, in medio dente armato. Long. Corp. 3¾ unc., (absque cornu capitis).
DYNASTES: coppery black; the thorax with three horns, the central one very short; the head with a long ascending horn, with a tooth in the middle. Length of the body, without the horn of the head, 3¾ inches.

SYN. Scarabæus Atlas? *Linn. Syst. Nat. vol.* I. *pars* 2, *p.* 542. *Mus. Lud. Reg.* 6. *Fabricius Syst. Eleuth.* I. *p* 10, *No.* 29. *Swammerdam's Book of Nature, tab.* 30, *f.* 3. *Margr. Brazil,* 247, *f.* 1. *Merian Ins. Surinam, in title page. Edwards's Birds, tab.* 105, *f.* 1.

This is one of the most extraordinary species of the great Scarabæi of Linnæus. The specimen here figured was purchased by the late Mr. Tunstal, from the cabinet of a Dutch governor in the East Indies, with other rare species figured in this work. At the period of the first publication of this work it was considered unique, and was regarded as an inhabitant of the island of Amboyna, in the East Indies. Linnæus, Fabricius, Madame Merian, and Margrave, give South America as the locality of the insect described in the works of the two former authors under the name of Scarabæus

B

Atlas, whilst Swammerdam and Edwards, who are both referred to by Linnæus, give Japan and Borneo as its habitat.

That the East Indies is the real locality of the insect here represented, is rendered most probable, from the circumstance of several other, and very closely allied species, being found in that part of the world. Such are the Scarabæus Caucasus, *Fabr.* (described from an East Indian specimen in the British Museum, and regarded by Olivier, and Jablonsky, as a variety of Atlas); Scarabeus Charon, *Oliv.*; the Javanese Scarabæus Hector, *Dejean;* Dynastes Hesperus, *Erichson,* from the island of Luzon; and Dynastes Jephthah, of *MacLeay,* (in the collection of the Entomological Society). It is true that the chief distinctions amongst these species are found in the size of the horns of the head and thorax, and in the teeth, or serration of the hinder part of the horn of the head; but it is equally true, that we are by no means furnished with sufficiently accurate data for ascertaining the extent of variation in the cornuted Scarabæi in this respect, and we are therefore by no means enabled to regard these as satisfactory species. Indeed, in the volume of the Naturalist's Library devoted to exotic Coleoptera, we have an original figure of an insect, with the name of Atlas attached, in which the horn of the head has a double series of serratures from the base to the apex, and which was brought from Rangoon; and in the Fabrician description this horn is stated to be tridentate, the anterior tooth being the strongest.

As it is, however, most probable that some, at least, of the above mentioned species are distinct, it is convenient to regard these insects as forming a distinct subgenus, characterized not only by the peculiar structure of the horns, and the polished surface of the body, but by the variations in the structure of the mouth. Mr. Hope has accordingly given to them the name of Chalcosoma, in his Coleopterous Manual, recently published.

ONTHOPHAGUS SPINIFEX.

Plate II. fig. 1.

FAMILY.   SCARABÆIDÆ, *MacLeay.*
GENUS.   ONTHOPHAGUS, *Latreille.* Copris p. *Fabricius.* Scarabæus p. *Linnæus.*
SPECIES. ONTHOPHAGUS SPINIFEX : thorace rotundato inermi ; occipite spinâ recurvâ thoracis longitudine ; thorace et elytris nigris, viridi nitidis ; his striatis ; pedibus nigris. Long. Corp. lin. $3\frac{1}{2}$.
ONTHOPHAGUS: with a rounded and unarmed thorax ; the forehead with a recurved spine as long as the thorax ; the latter, as well as the elytra, black shaded with green ; the elytra striated, and the legs black. Length $3\frac{1}{2}$ lines.

COLEOPTERA.

SYN. Scarabæus spinifex, *Fabricius Ent. Syst.* 1. *p.* 58. *sp.* 190. *Herbst. Coleopt.* II. *p.* 240. *No.* 146. *Donovan edit.* 1, *in text.*
Scarabæus spinifer, *Oliv. Ent.* 1. 3. 148. *No.* 180. *t.* 12. *f.* 112. *Donovan edit.* 1. *in tab.*

Fabricius describes this species as being of the same size as Onthophagus nuchicornis, with the occipital horn as long as the thorax, and with the thorax and elytra black tinged with green. The specimen described by Fabricius, was brought from the Coast of Coromandel by Sir Joseph Banks. Donovan's specimens were from Bengal.

GYMNOPLEURUS MILIARIS.

Plate II. fig. 2.

GENUS. GYMNOPLEURUS, *Illiger.* Ateuchus p. *Fabricius.*
SPECIES. GYMNOPLEURUS MILIARIS: clypeo 6-dentato; thorace et elytris obscuris, maculis elevatis atris nitidulis. Long. Corp. lin. 4.
GYMNOPLEURUS: with the clypeus 6-dentate; the thorax and elytra obscurely coloured, with shining elevated black spots. Length of the body, one-third of an inch.
SYN. Scarabæus miliaris, *Oliv. Ins.* 1. 3. 167. 206. *t.* 18. *f.* 164. *Fabricius Ent. Syst.* 1. *p.* 63. *No.* 209. *Herbst. Col.* II. *p.* 322. *No.* 206.
Ateuchus miliaris, *Fabr. Syst. Eleuth. vol.* 1. *p.* 56.
HABITAT. "India," (*Fabricius*).

This and the following species seem at first sight to resemble each other, but are nevertheless distinct: they are both represented in the same plate, of the natural size and magnified, in order that the exact difference between them may be the more easily discriminated. The pale colour of the thorax and elytra is produced by numerous minute incumbent hairs, of which the elevated tubercles are destitute.

GYMNOPLEURUS KŒNIGII.

Plate II. fig. 3.

SPECIES. GYMNOPLEURUS KŒNIGII: clypeo bidentato; thorace et elytris pallidis, maculis atris elevatis nitidulis, his variolosis. Long. Corp. lin. 4.
GYMNOPLEURUS: with the clypeus bidentate; the thorax and elytra pale, with elevated shining black spots, the latter variolose, with some of the spots confluent. Length of the body, 4 lines.

COLEOPTERA.

Syn. Scarabæus Kœnigii, *Oliv. Ins.* 1. 3. 163. 200. *t.* 9, *f.* 77 *Fabricius Ent. Syst.* 1, *p.* 65.
Herbst. Col. II. *p.* 300, *No.* 195, *t.* 19, *f.* 8.
Scarabæus scriptus, *Pallas Icon. p.* 7. *A.* 7. *T. A. f.* 7.

Fabricius gives Tranquebar as the locality of this species. I have seen it in some collections named Gymnopleurus granulatus, which Fabricius, indeed, regarded as a variety of Gymnopleurus Kœnigii.

CETONIA HISTRIO.

Plate II. fig. 4.

Family. CETONIADÆ, *MacLeay.*
Genus. CETONIA, *Fabricius.* Scarabæus p. *Linnæus.*
Species. CETONIA HISTRIO : testacea ; thorace maculis duabus nigris ; elytris nigris, singulo macula magna oblonga testacea, punctisque quatuor lateralibus albis. Long. Corp. lin. 8.
CETONIA : testaceous ; thorax with two black spots ; elytra black, each having a large oblong testaceous spot, and four lateral small round white spots. Length 8 lines.
Syn. Cetonia histrio, *Oliv. Hist. Ins.* 1. 6. 45, *tab.* 10, *f.* 94. *Fabricius Ent. Syst.* 2. 152.
Cetonia versicolor, *Oliv. Fabr. var. ? Schonh. Syn. Ins.* 1. 3, *p.* 131.
Habitat. Egypt, *(Fabricius).* Egypt, and East Indies, *(Schonherr).* Bengal, *(Mus. Hope).*

GYMNETIS CÆRULEA.

Plate II. fig. 5.

Genus. GYMNETIS, *MacLeay.* Cetonia p. *Fabricius.*
Species. GYMNETIS CÆRULEA : glaberrima, glauca, elytris chalybæis, punctato striatis, maculis albis. Long. Corp. lin. 5½.
GYMNETIS : very smooth, glaucous, with the elytra chalybeous, striato punctate, and spotted with white. Length 5½ lines.
Syn. Scarabæus cæruleus, *Herbst. in Fuessly Archiv Ins.* 4, *p.* 19, *No.* 8, *t.* 19, *f.* 30. *Gmel. Syst. Nat.* 1582, *No.* 382. *Oliv. Ent.* 1. 6, *p.* 47, *t.* 5, *f.* 31, *a.*
Cetonia 14-maculata, *Fab. Syst. Eleuth.* 2, *p.* 156.
Habitat. India, *(Olivier).* Bengal, *(Donovan).*

The original specimen of Cetonia 14-maculata, described by Fabricius, is now in the collection of the Rev. T. W. Hope, an examination of which has convinced

COLEOPTERA.

me that it is identical with the insect here represented. There are two oblique white spots at the base of the suture of the elytra, which gives the insect the appearance of possessing a distinct scutellum. Olivier, however, correctly described the thorax as being posteriorly lobed, which is the true character of the genus Gymnetis.

BUPRESTIS (STERNOCERA) STERNICORNIS.

Plate III. fig. 1.

TRIBE. STERNOXI, *Latreille.*
FAMILY. BUPRESTIDÆ, *Leach.*
GENUS. BUPRESTIS, *Linnæus.* (SUBGENUS: Sternocera, *Eschscholtz, Solier.*)
SPECIES. BUPRESTIS STERNICORNIS: inauratus, punctis cinereis impressis; elytris serrato-tridentatis; sterno porrecto conico. Long. Corp. $1\frac{1}{2}$ unc.
BUPRESTIS: shining green, with grey impressed spots; elytra serrated, and tridentate at the tips; sternum porrected and conical. Length of the body $1\frac{1}{2}$ inch.
SYN. Buprestis sternicornis, *Linn. Syst. Nat.* I. II: *p.* 610. *Fabricius Ent. Syst.* I. II. *p.* 194. *Oliv. Ent.* II. 32. *p.* 35. *t.* 6. *f.* 52 *a.* *Herbst. Col.* IX. *p.* 16. 2 *t.* 138. *f.* 3.

Among the insects of China is a beautiful species of the genus Buprestis, which the natives of that country collect in considerable numbers, and employ in the various ornaments of their dresses, arms, &c. The Buprestis sternicornis, and chrysis, are collected in India for similar purposes, but being scarce, are esteemed more valuable than the other kind, which they receive at a low price from China.

Both species are brought from Madras and Bombay, but generally in a mutilated state; for the Indians perforate them at both ends, and string them like beads, when they collect them.

BUPRESTIS (STERNOCERA) CHRYSIS.

Plate III. fig. 2.

SPECIES. BUPRESTIS CHRYSIS: thorace viridi; elytris serrato-tridentatis, castaneis lævibus; sterno conico porrecto. Long. Corp. unc. $1\frac{1}{2}$.
BUPRESTIS: with the thorax shining green; the elytra chestnut-coloured, smooth, and terminated by three teeth; sternum conical and porrected. Length $1\frac{1}{4}$ inch.

COLEOPTERA.

SYN. Buprestis chrysis, *Fabricius Ent. Syst.* 1. 11. p. 194. *Oliv. Ent.* II. 32, t. 2, f. 8, a, d, e. *Herbst. Col.* IX. p. 14, 1 t. 138, f. 2.
Buprestis sternicornis, *De Geer Ins.* IV. p. 136. 2, t. 17, f. 25.

This and the preceding species are alike distinguished by the comparatively small size of the male sex, in which the fifth ventral segment is truncated at its extremity, whereas in the females it is rounded.

BUPRESTIS (SPHENOPTERA ?) CONFUSA.

Plate III. fig. 3.

SPECIES. BUPRESTIS CONFUSA : corpore æneo immaculato, obscure nitido ; elytris striatis et tridentatis. Long. Corp. lin. 6.
BUPRESTIS : brassy coloured, without spots, obscurely shining, with the elytra striated and tridentate at the tips. Length $\frac{1}{2}$ inch.

SYN. Buprestis confusa, *Westwood*.
Buprestis ænea, *Fabricius Ent. Syst.* 1, 11. p. 193. (*nec ænea, Linnæi.*) *Syst. Eleuth.* 2. p. 193. *Donov. 1st edit.*
Buprestis tricuspidata, *Oliv. Ent.* 11. 32. p. 29. t. 8. f. 87. *Herbst. Col.* IX. p. 181. t. 154. f. 1. a, b ?

The original specimen described by Fabricius, was brought from Coromandel by Sir J. Banks. The name applied to it by the former having been previously given by Linnæus to another species, from the south of Europe, (Buprestis, Diceræa, ænea,) it has become necessary to substitute another in its stead, as it is evident that it is not identical with the Buprestis tricuspidata of Olivier, which, according to Schonherr, is from Sierra Leone.

BUPRESTIS (ANTHAXIA) 4-MACULATA.

Plate III. fig. 4.

SPECIES. BUPRESTIS 4-MACULATA : elytris integris, viridis ; thorace posticè late elytrisque maculis duabus magnis aureis. Long. Corp. lin. $4\frac{1}{2}$.
BUPRESTIS : with the elytra entire ; green ; hinder part of the thorax with a broad golden band ; elytra with two large golden spots on each Length $4\frac{1}{2}$ lines.

COLEOPTERA.

Syn. Buprestis 4-maculata, *Fabr. Sp. Ins.* I. *p.* 280. *Syst. Eleuth.* 2. *p.* 208. *Olivier Ent.* II. *p.* 32. *t.* 10. *f.* 110. *Herbst. Col.* IX. *t.* 153, *f.* 4.

This admirably beautiful species is described by Fabricius as an inhabitant of India, on the authority of the cabinet of the late Dr. Fothergill. It bears very great resemblance to several brilliant species found in South America and the West Indies, composing the subgenus Actenodes of Dejean. Schonherr, most probably by mistake, has given Italy as the locality of this species.

ANTHIA 6-GUTTATA.

Plate IV. fig. 1.

Tribe. Geodephaga, *MacLeay.*
Family. Carabidæ.
Genus. Anthia, *Weber.*
Species. Anthia 6-Guttata : atra ; elytris lævibus ; thorace maculis duabus coleoptrisque quatuor albo tomentosis. Long. Corp. 17—21 lin.
Anthia : black, with smooth elytra ; thorax with two, and elytra with four white downy spots. Length of the body 17 to 21 lines.
Syn. Carabus sex-guttatus, *Thunberg nov. Sp.* 3. 70. *fig.* 84. *Gmel. Syst. Nat.* 1965. *Fabr. Ent. Syst.* 1. 141. 75. *Syst. Eleuth.* 1. *p.* 221. (Anthia 6-guttata.)
Carabus sex-maculatus, *Donov. edit.* 1. (nec *Fabr. Ent. Syst.* 1. 141. 78.) *Syst. Eleuth.* 1. 222.

This conspicuous insect is one of the largest species of the Linnæan genus Carabus. It is very abundant in some parts of India. Donovan, although referring to the original description of A. sex-guttata, published this under the name of Carabus 6-maculatus, which name had been given by Fabricius to a different species of the same genus.

BRACHINUS BIMACULATUS.

Plate IV. fig. 2.

Genus. Brachinus, *Weber.* Carabus p. *Linnæus, Fabricius.*
Species. Brachinus Bimaculatus : capite flavescente ; vertice obscuro ; thorace obscuro, maculis duabus flavescentibus ; elytris nigris, puncto humerali, fascia lata media sinuata abbreviata, apice, antennis pedibusque flavescentibus. Long. Corp. lin. 7—8.

COLEOPTERA.

BRACHINUS: with the head yellow; the crown dark coloured; the thorax dark, with two large yellow spots; the elytra black, with a humeral spot; a broad central fascia, interrupted in the middle, and the tips, as well as the antennæ and legs, yellowish. Length 7 to 8 lines.

SYN. Brachinus bimaculatus, *Dej. Spec. gen.* 1. *p.* 299.   *Oliv.* III. 35. *p.* 65. *No.* 81. *t.* 2. *fig. a, b, c. Schonherr Syn. Ins.* 1. *p.* 229. *No.* 1.   (G. Pheropsophus, *Solier.*)

I have not cited Fabricius, nor the other old authorities, as it is evident from their descriptions that they confound several distinct, but nearly allied, species together. Donovan's figure agrees, however, with Dejean's Brachinus bimaculatus, which is from the East Indies.

PLATYRHOPALUS DENTICORNIS.

Plate V. fig. 1.

TRIBE. XYLOPHAGA?
FAMILY. PAUSSIDÆ, *Westwood.*
GENUS. PLATYRHOPALUS, *Westwood.*   Paussus p. *Donovan.*
SPECIES. PLATYRHOPALUS DENTICORNIS: brunneo-rufescens; elytris dorso fuscis; suturâ latè ad basin maculâque utrinque posticè, rufescentibus; antennarum clavâ magnâ latere omni acuto, juxta basin externe incisâ.   Long. Corp. lin. 4—5.
PLATYRHOPALUS: brownish-red, with the middle of the elytra brown; a broad mark at the base of the suture, and two spots near the extremity red; the club of the antennæ large, ovate, margin acute, with a deep notch near the base behind. Length 4 to 5 lines.

SYN. Platyrhopalus denticornis, *Westwood in Trans. Linn. Soc. vol.* XVI. *p.* 657.
Paussus denticornis, *Donovan, 1st edit.*

HABITAT. Bengal.

The first account of the genus Paussus appears in a small tract written by Linnæus, and published at Upsal, in the year 1775, under the title of *Bigæ Insectorum,* &c. This paper contains likewise a description of the Diopsis genus, which, together with the Paussus, are unquestionably two of the most singular genera of the many tribes of insects hitherto discovered. Both may possibly derive some additional celebrity also, from the recollection that the dissertation in which they are inserted, concluded the Entomological labours of that distinguished naturalist: it was the last he ever published in the department of zoology.

COLEOPTERA.

In the dissertation alluded to, the genus Paussus is exemplified by a single species, P. microcephalus, and Diopsis by D. ichneumonea, a plate with figures of both of which, drawn by J. Afzelius, and engraved by Berquist, accompany the descriptions. It is to this plate, and the original descriptions of Linnæus, that Fuessly is indebted solely for the account he gives of both these genera, in his *Archiv. der Insectengeschichte*, printed at Zurich in 1783, as well as in the French translation of that work which afterwards appeared in Paris. Indeed, as Professor Afzelius has suggested, from the repeated errors that appear in those works, in translating the Linnæan observations, defining the character of the genus Paussus, &c., it is very likely that neither Fuessly, nor his translators, Herbst, Gmelin, and some other writers who have treated on it, ever saw an insect either of this genus, or of Diopsis.

Thunberg, during his travels through the country of the Hottentots in 1772, found two coleopterous insects, which he conceived, with much propriety, ought to be referred to a new genus, none of those established previous to his departure from Europe by Linnæus being calculated to admit them. But on his return to Sweden, he found that Linnæus in his absence had described the genus Paussus, to which they might be referred. An account of these was afterwards inserted in the Transactions of the Royal Academy of Stockholm for 1781: this paper is accompanied with a figure of only one of the insects mentioned, P. lineatus, a species very aptly named, from the distinct longitudinal streak on each of the wing cases; the other insect described by Thunberg, he calls ruber. Fabricius consigned these, with the Linnæan insect, to his genus Cerocoma.

The latest history of the genus Paussus, previous to the first edition of this work, was from the pen of Professor Afzelius, a learned, copious, and elaborate paper, inserted in the fourth volume of the Transactions of the Linnæan Society, in which he describes Paussus microcephalus, and another species which he found in Africa, which he names P. sphærocerus. Neither of those insects are allied to the species figured in this plate, which were entirely undescribed. For this important accession of new species to a group so little known, Donovan was indebted to the active and praiseworthy zeal of Mr. Fichtel, in compliment to whom one of them is named Fichtelii.

Since the publication of the first edition of this work, various additional species have been discovered, constituting a very natural family, distributed into several distinct genera, of which I have published a monograph in the sixteenth volume of the Linnæan Transactions. I have also more recently become acquainted with several other species, of which figures and descriptions have been laid before the Entomological Society of London.

COLEOPTERA.

Donovan, in order to admit this species into the genus Paussus, was induced to adopt the Linnæan generic character, with the omission of the term "clava solida," although, according to the characters laid down by Afzelius, it ought not to be admitted into the genus, since he states that the tarsi are only three-jointed, whereas in the other species they are five-jointed. They are however four jointed, but in consequence of the more important variation in the structure of the mouth, which Donovan neglected to examine, I considered it requisite to establish this species as a distinct genus.

PAUSSUS THORACICUS.

Plate V. fig. 2.

GENUS. PAUSSUS, *Linnæus.* Pausus, *Donovan.*

SPECIES. PAUSSUS THORACICUS: ferrugineo-testaceus; elytris disco lateribusque fuscis: antennarum clava oblongâ compressâ trigonâ; latere interno acuto, externo excavato; cavitate ovali marginibus denticulatis. Long. Corp. lin. 3⅓.

PAUSSUS: pale reddish; the elytra having the disc and sides brown; the club of the antennæ oblong, compressed, and triangular; the inner margin acute, the exterior excavated, with an oval cavity, the margins being denticulated. Length 3⅓ lines.

SYN. Paussus thoracicus, *Westwood in Trans. Linn. Soc. vol.* XVI. *p.* 640, *pl.* 33, *f.* 28—30.
Pausus thoracicus, *Donovan Ins. Ind.* 1*st edit.*
Paussus trigonicornis, *Latr. Gen. Cr. &c.* 3. *p.* 3, *pl.* 11, *f.* 8. *Sch. Syn. Ins.* 1. *p.* 3, *pl.* 9.

HABITAT. Bengal.

The thorax in this insect is so deeply divided across the middle, that it appears, at first sight, as if it were really two, whence the specific name thoracicus. The same character also exists in P. Fichtelii, which Donovan considered might possibly be the other sex of P. thoracicus; the most striking difference prevailing in the structure of the antennæ, the excavation in one of which is of an oval, or rather shuttle shape, and in the other pyriform. In my monograph I have pointed out other characters which have induced me to consider the two species as entirely distinct.

COLEOPTERA.

PAUSSUS FICHTELII.

Plate V. fig. 3.

SPECIES. PAUSSUS FICHTELII: testaceus; elytris fuscis, lateribus basi apiceque testaceis; thorace subbipartito; antennarum clavâ oblongâ, latere interno acuto, externo excavato, cavitate pyriformi, marginibus denticulatis. Long. Corp. lin. 2½.

PAUSSUS: pale reddish, with brown elytra, having the sides, base and tips reddish; thorax nearly divided into two parts; club of the antennæ oblong, with the inner margin acute, the external excavated with a pear-shaped cavity, the margins being denticulated. Length 2½ lines.

SYN. Paussus Fichtelii, *Westwood in Trans. Linn. Soc. vol.* XVI. *p.* 641, *pl.* 33, *f.* 31—33.
Pausus Fichtelii, *Donovan Ins. Ind.* 1st edit.

HABITAT. Bengal.

There is a specimen of this species in the collection of the Entomological Society of London, presented by the Rev. W. Kirby.

PAUSSUS PILICORNIS.

Plate V. fig. 4.

SPECIES. PAUSSUS PILICORNIS: testaceus; elytris piceis: thorace bipartito; antennarum clavâ oblongâ, apice attenuatâ incurvâ, pilis longis sparsis. Long. Corp. lin. 2.

PAUSSUS: reddish, with pitchy elytra; the thorax nearly divided into two parts; club of the antennæ oblong, attenuated, and incurved at the tip, with long hairs. Length 2 lines

SYN. Paussus pilicornis, *Westwood in Trans. Linn. Soc. vol.* XVI. *p.* 643.
Pausus pilicornis, *Donovan Ins. Ind.* 1st edit.

HABITAT. Bengal.

In the formation of the club composing the second or exterior joint of the antennæ, this species differs altogether from the others: the club is entire or not excavated, and is slightly beset with hairs. Of this species Mr. Fichtel met with only a solitary specimen, as was likewise the case with Paussus Fichtelii.

COLEOPTERA.

PREPODES REGALIS.
Plate VI. fig. 1.

TRIBE. RHYNCOPHORA, *Latreille.*
FAMILY. CURCULIONIDÆ, *Leach.*
DIVISION. BRACHYDERIDES, *Schonherr.*
GENUS. PREPODES, *Schonherr.* Circulio, *Linn. &c.*
SPECIES. PREPODES REGALIS: oblongo ellipticus, niger, squamulis cœruleo-virentibus tectus, rostro subcarinato; thorace supra impresso vitta media lateribusque cupreo-aureis; elytris subtilius remote punctato-striatis, plagâ baseos, fasciisque tribus flexuosis cupreo-aureis nigro-marginatis ornatis. Long. Corp. lin. 8.

PREPODES: oblong-elliptical, black, clothed with greenish-blue scales; the thorax with a central streak, and the sides golden coppery; the elytra with a basal spot and three waved bands, coppery golden coloured, margined with black. Length 8 lines.

SYN. Curculio regalis, *Linn. Syst. Nat.* I. II. p. 616. *Fabr. Syst. Eleuth.* 2. 508. *Oliv. Ent.* V. 83, t. 1. f. 8, a, b.
Prepodes (Callizonus, *Olim.*) regalis, *Sch. Sp. Curcul.* 2. p. 21.

Donovan well observed that this species truly deserves the title regalis, being an aggregate of beauty and splendour: of the loveliest cœrulean, changing alternately to the deeper glow of the violet, to green, or the transitory sparkling of intermingled silver. Every space of blue is constantly contrasted with another of crimson, and which, as the violet changes to blue or green, alters its aspect to a still more vivid expanse of gold. Each of those colours, the blue and red, are distinct; for an irregular space of black limits every spot and marking, and relieves the whole. As the effect of such a combination of colours in this comparatively small species is inconceivably splendid, and almost inimitable, one figure in the plate is intended to show the natural size, and to admit of more perfect delineation, another somewhat magnified is added also.

This insect, which Donovan believed to be *unique* in Great Britain, was brought from France, in the collection of Mons. De Calone, and was in the possession of the author. Linnæus met with it in one of the cabinets on the continent, and described it as a South American insect. In the works of Fabricius, its habitat is given " in *Indiis*." On this authority Donovan introduced it amongst the insects of *East* India, but it was the practice of Fabricius, and the older authors, to employ the term " in Indiis," to designate the *West* as well as the *East* Indies. The real locality of this species is South America, Peru, and the island of Saint Domingo.

COLEOPTERA.

CALANDRA PALMARUM.

Plate VI. fig. 2.

DIVISION. RHYNCOPHORIDES, *Schonherr*.

GENUS. CALANDRA, *Fabricius*. Rhyncophorus, *Herbst*. Curculio p. *Linn*.

SPECIES. CALANDRA PALMARUM : depressa, atra, obscura ; thorace supra plano ; elytris abbreviatis striatis ; rostro apice compresso, supra piloso. Long. Corp. 1⅜ unc.
CALANDRA : depressed, black, not shining ; thorax flat above ; elytra abbreviated and striated ; rostrum compressed at the tip, and hairy above. Length 1⅜ inch.

SYN. Curculio palmarum, *Linn. Syst. Nat.* 1. 11. p. 606. 1. *Fabricius Syst. Eleuth.* 2. 430. De Geer *Ins.* 5. t. 15. f. 26.

Donovan observes, upon this insect,—" A very abundant species in India, where it is found chiefly on the palm tree." The real locality of this species is South America. De Geer and Madame Merian giving Surinam; Dejean, Cayenne; and Latreille, South America, Cayenne, and Surinam. The mistake appears to have originated, either by confusing several allied species together, or by giving the West Indies under the name of India generally, as noticed under the last species.

Order. HEMIPTERA, *Linnæus*.

FULGORA PYRORHINA.

Plate VII. fig. 1.

SUBORDER. HOMOPTERA, *Latreille*.
FAMILY. FULGORIDÆ, *Leach*.
GENUS. FUGORA, *Linnæus*.
SPECIES. FULGORA PYRORHINA : "fronte rostrata adscendente, apice rubra ; hemelytris fuscis fascia pallidiori ; alis nigris, basi viridibus." *Donovan*. Expans. alar. unc. 4.
FULGORA : with the trunk rostrated, ascending ; red at the tip ; wing-cases brown, pale across the middle ; wings black, green at the base. Expanse of the wings 4 inches.

SYN. Fulgora pyrorhina, *Westwood Mon. Fulgorained.*
Fulgora pyrorhynchus, *Donovan*, 1st edit.

" Amongst the more valuable acquisitions, designed to enrich this illustration, few can afford higher gratification to the scientific reader than this Fulgora. In size it is inferior only to F. lanternaria ; it is an undoubted nondescript, and may be con-

HEMIPTERA.

sidered as a striking example of the entomological riches of a country hitherto scarcely known, the interior of Indostan. It was originally brought from India by the late Governor Holford, and is now in the possession of the author. He has sought in vain for this species in other cabinets of exotic insects, and ventures to deem his specimen unique."

Donovan, adopting the opinion of the luminosity of the Fulgoræ, then proceeds: —" In the course of our remarks on the Fulgora so abundant in China, F. candelaria, our attention was naturally directed to the astonishing property some insects of this genus are known to possess, that of emanating light; and it was to this insect we alluded in particular, when speaking of one from interior India, that enabled us to extend our observations on that property. The trunk is large, of a dark purple, thickly sprinkled with spots of white phosphoric powder, and the apex, which is scarlet, and somewhat pellucid, still retains a reddish glow, that almost convinces us the creature when living could diffuse light both from the apex and the spots. In admitting this conjecture, without wandering into the marvellous, its nocturnal appearance must be infinitely more singular than either of the known species of Fulgoræ, Lampyrides, or any other luminous insect yet discovered, for, when on the wing, the illuminated apex would resemble a globule of fire, or heated iron, and the numberless phosphoric spots on the tube form a train of glittering stars to accompany it.*

" The only figure of a Fulgora in any respect resembling this species is given in the works of Stoll, under the title of *De Groote Groene Coromandelische Lantaarndrager;*† but among other evident specific distinctions we need only notice the structure of the trunk, which is altogether different, being much recurved, and tapering gradually from the base to an acute point at the apex: its colour is also an olive black."

---

* " As it may be thought improbable that any insect can exhibit such an extraordinary appearance, the words of Olivier on some species of Lampyrides may not be unsatisfactory. The insects are certainly very distinct, but reports of travellers countenance an opinion that the phosphoric emanations are analogous in the species of both genera. 'The phenomena produced by a natural phosphorus is still more wonderful in some foreign species, in which the males shine, and being provided with wings, will produce in their rapid flight a thousand small stars.'—*Olivier, Histoire des Insectes.*"

† " Pl. 26, fig. 143. Green lantern-carrier fly of Coromandel.—At the conclusion of the description, Stoll says, '*In de Nederlandsche Kabinetten, &c. &c.*'—' This insect was not known in the cabinets of the low countries till within three years, [anno 1780,] during which time a few were brought from Tranquebar, on the Coromandel coast, to the cabinet of natural curiosities of his Highness the Stadtholder of the United Provinces, of which I have been obligingly permitted to take the figure of a female by Mons. Vosmar, to whom I owe my public acknowledgments for it.'—99 *tab.* 26."

HEMIPTERA.

APHANA? FESTIVA.

Plate VII. fig. 2.

GENUS. APHANA? *Burmeister.* Aphæna, *Guerin.*

SPECIES. APHANA? FESTIVA : fronte conica ; hemelytris fuscis margine antice virescente punctis 5 atris, quatuor postici inter puncto fulvo ; alis sanguineis apice fuscis. Long. (alis clausis) ¾ unc.

APHANA? with the trunk short, conical ; wing-cases brown, with the front margin greenish, in which are five black spots, the four exterior ones having a fulvous dot ; wings sanguineous, with the tips brown. Length with the wings closed ¾ of an inch.

SYN. Fulgora festiva, *Fabricius, Ent. Syst.* 4. *p.* 5. *Syst. Rhyng, p.* 4.

HABITAT. " Coromandel, *Mus. Dom. Banks,*" *Fabricius.*

In addition to the Banksian specimens, still preserved at the Linnæan Society, there is a specimen of this curious species in the collection of the Entomlogical Society of London, from the Kirbian cabinet. The antennæ being mutilated, I am unable to decide whether it be strictly an Aphana, to which genus, however, it is most nearly allied.

PSEUDAPHANA HYALINATA.

Plate VII. fig. 3.

GENUS. PSEUDAPHANA, *Burmeister.* Dictyophora, *Germar.* Fulgora p. *Fabr.*

SPECIES. PSEUDAPHANA HYALINATA : fronte conica inæquale ; hemelytris hyalinis nervis albo nigroque punctatis striaque transversa in medio atra ; alis hyalinis, macula apicis atra. Long. alis cl. unc. 1.

PSEUDAPHANA : with the trunk short, conical, uneven above ; wing-covers hyaline, with the veins spotted alternately with black and white, and with a transverse black central band ; wings hyaline with a black apical spot. Length with the wings shut, 1 inch.

SYN. Fulgora hyalinata, *Fabr. Ent. Syst.* 4. *p.* 5. *Syst. Rhyng. p.* 4.

Pseudalphana h., *Burmeister Handb. der Ent.* 2. *p.* 160. *Germar in Thon's Arch.* 2. *p.* 47.

Fabricius refers to the collection of Sir Joseph Banks, Bart. for this and the preceding species. The annexed figures are copied from the specimens that author has described.—F. Hyalinata is from Bengal.

HEMIPTERA.

FULGORA LINEATA.

Plate VIII. fig. 1.

SPECIES. FULGORA LINEATA: fronte rostrata lineari adscendente; hemelytris pallidis, lineis duabus fuscis. Long. alis clausis. unc. ½.

FULGORA: with the trunk linear ascending; hemelytra pale, with two brown longitudinal lines. Length with the wings closed, ½ an inch.

"A pretty little undescribed species, found in Bengal, where it is not very uncommon."—*Donovan*.

An apparent variety of this species is figured in *Griffith's Animal Kingdom*,—*Insects, pl.* 90, *fig.* 2, from India, under the name of Fulgora pallida, a name which, at all events, must be rejected, having been previously given to the following species.

PSEUDAPHANA PALLIDA.

Plate VIII. fig. 2.

SPECIES. PSEUDAPHANA PALLIDA: "fronte rostrata lineari, adscendente; thorace pallide viridi, rubro lineato; hemelytris hyalinis."—*Donovan*. Long. alis clausis, ½ unc.

PSEUDAPHANA: with the trunk linear, ascending; thorax pale green with red lineations; wing-cases hyaline. Length with the wings closed, ½ an inch.

SYN. Fulgora pallida, *Donovan, 1st edit.*

HABITAT. "From the same place as the preceding insect."—*Donovan*.

The small species of Pseudaphana, which is well distinguished from Fulgora by the less closely reticulated wings, are somewhat numerous. Three species have been described as inhabitants of the East Indies, namely, Fulgora grammea, *Fabricius* (*Syst. Rhyng.* p. 4 "*India Orientali.*"); Flata lyrata, *Germar*, (*Thon's Ent. Archiv.* 2. p. 47, "*Bengalia.*"); and Flata splendens, *Germar* (*Thon's Ent. Archiv.* 2. p. 47, "*Java.*"). It is not improbable that Donovan's species may be identical with one of these, possibly the F. lyrata, which is also from Bengal; but from the too concise description of Donovan, which I have copied entire above, it is not possible to speak with absolute precision.

HEMIPTERA.

CICADA SPECIOSA.

Plate VIII. fig. 3.

FAMILY. CICADIDÆ, *Leach.*
GENUS. CICADA, *Linnæus.* Tettigonia, *Fabricius.*
SPECIES. CICADA SPECIOSA: nigra ; thorace fascia lata flava ; abdomine postice fascia aurantia ; hemelytris fusco-olivaceis venis rubris ; alis margine albo. Expans. alar. unc. 6¼.

CICADA : black ; thorax with a broad yellow fascia ; abdomen with the fifth, sixth and seventh segments fulvous above ; hemelytra brown-olive with red veins ; wings with the margin white. Expanse of the wings, 6¼ inches.

SYN. Tettigonia speciosa, *Illiger in Wied. Archiv. fur. Zool. and Zoot.* 2. 145. 38. tab. 2. (1800). *Fabr. Syst. Rhyng, p.* 33.
Cicada Indica, *Donovan, 1st edit.*

"This is unquestionably one of the most striking and magnificent species of Cicada we are acquainted with. A single specimen of this kind was discovered in Bengal by Mr. Fichtel about four years ago. It is now deposited in the Imperial cabinet at Vienna." (*Donovan.*)

The species was also contained in the collection of Hellwig, from which it was described and figured by Illiger, in the work above referred to, in the year 1800, for which reason I have retained the latter name as having the priority.

Order. **ORTHOPTERA**, *Olivier.* (**HEMIPTERA** p. *Linnæus.*)

PHASMA (CYPHOCRANA) GIGAS.

Plate IX.

SECTION. CURSORIA, *Latreille.*
FAMILY. PHASMIDÆ, *Kirby.*
GENUS. PHASMA, *Fabricius.*
SUBGENUS. Cyphrocana, *Serville.* Mantis, *Linn. Fabr.* Phasma, *Stoll. Fabr.*

ORTHOPTERA.

SPECIES. PHASMA (CYPHOCRANA) GIGAS : viridis; mesothorace scabro ; tegminibus ovalibus; alis (area costali viridi,) obscuré fulvis fusco undatis ; pedibus spinosis, flavoridibus. Long. Corp. unc. 7. Expans. alar. unc 7.

PHASMA (CYPHOCRANA) : green ; mesothorax scabrous; wing-covers small oval ; wings, (with the costal area green,) dirty fulvous with brown waves ; legs spinose, yellowish-green. Length 7 inches. Expanse of the wings 7 inches.

SYN. Mantis Gigas, *Linn Syst. Nat.* 2. 689. *Fab. Ent. Syst. vol.* 2. *p.* 14. *Ent. Syst. Suppl. p.* 187. *Shaw. Miscell. pl.* 43. *Donovan, 1st edit. (nec* Phasma (Diapherodes) Gigas, *Drury,* 2. *pl.* 50.)

Cyphocrana Gigas, *Serville Ann. Sc. Nat.* XXII. *p.* 60. *Gray Synops. Phasm. p.* 35. *Le Geant. Stoll Spectres. pl.* 2, *fig.* 5.

Donovan observed of this insect, that it was the largest species of this very extraordinary genus known, and that this specimen was from the island of Amboyna, where it was rare. It is, however, exceeded in size by the Phasma (Diura) Titan of MacLeay, figured by Gray, in his beautiful Memoir of the Phasmidæ of New Holland, plate 4.

PHASMA (PLATYCRANA) EDULE ♀.

Plate X.

SUBGENUS. PLATYCRANA, *Gray.* CYPHOCRANA, *Serville.*

SPECIES. PHASMA (PLATYCRANA) EDULE : viridis ; mesothorace tereti ; ♂ scabro, ♀ glabro ; tegminibus area costali viridi, basi coccineis ; pedibus brevibus, submuticis ; femoribus posticis spinosis. Long. Corp. ♀ unc. 6. Expans. alar. unc. 4½.

PHASMA (PLATYCRANA) : green ; with the mesothorax slender ; scabrous in the male, smooth in the female ; wing-covers with the costal area green, pink at the base ; legs short, almost smooth ; posterior femora spined. Length of the body ♀ 6 inches. Expanse of the wings 4½ inches.

SYN. Phasma edule, *Licht. in Linn. Trans. tom.* 6. *p.* 13.

Mantis viridis, *Donovan, 1st edit.*

Le Spectre verd, *Stoll, pl.* 6, *fig.* 20 ♀, 21 ♂.

Mantis viridana, *Oliv. Enc. Meth.* 7. 636. *Serville Revis. Orthopt. p.* 33. (*Ann. Sc. Nat.* 20. *p.* 60.) *Gray Synops. Phasm. p.* 36. (Platycrana v.)

Donovan states that he received this species also from Amboyna, where it is, perhaps, more scarce than the preceding insect.

ORTHOPTERA.

PHYLLIUM SICCIFOLIUM.

Plate XI. fig. 1. (Pupa, fig. 2.)

GENUS. PHYLLIUM, *Illiger, Latr.* Mantis, *Fabr. Donovan.* Pteropus, *Thunberg.*
SPECIES. PHYLLIUM SICCIFOLIUM : viride ; capite, thorace, pedibusque luteo-viridibus ; femoribus ovatis membranaceis ; tibiis quatuor posticis inermibus ; thorace denticulato. Long. Corp. 3½ unc.

PHYLLIUM : green ; with the head, thorax, and legs tinged with luteous ; the femora ovate, membranaceous ; the four posterior tibiæ not dilated ; thorax toothed at the sides. Length of the body, 3½ inches.

SYN. Mantis siccifolia, *Linn. Mus. Lud. Reg.* III. *Syst. Nat.* 2. p. 689. *Fabr. Ent. Syst.* 2. p. 18. *Donovan*, 1st edition.
La Feuille de citron, *Stoll. Spec. pl.* 7, ♂, ♀, *larva et pupa* (antennis ♀ et pupæ ? vitiosis)
Phasma citrifolium, *Lichtenstein Trans. Soc. Linn. t.* 6, p. 17
Phyllium (1) brevicorne. *Latr. Gen.* 3, p. 39, ♀. (Phyllium, sect. 11, ♂ ibid.)
Phyllium siccifolium, *Lat. R. An.* (2d edit.) 5. 179. (*Edit. Crochard*) *Ins. Livr.* I. pl. 79, *fig.* 1. *(with details, fig.* 16, *pupa maris nec imago* ♂ *). Serville Revis. Orthopt.* p. 36. (*Ann. Sc. Nat.* 22. 63.) *Gray Synopsis Phasm.* p. 30.

"An erroneous opinion has prevailed pretty generally among naturalists respecting the colour of this insect, which when living they conceived to be similar to that of a dried or withered leaf. This, it may be observed, is commonly the appearance of the insect after death : such was no doubt the colour of the specimens delineated by Roesel ; nor can we for a moment hesitate in believing that the insects described by the accurate Linnæus and Dr. Shaw exhibited the like appearance. The specimen of the winged insect in our cabinet, has been preserved, however, with more than usual care. Immediately after the death of the creature, as we have reason to suspect, the abdomen had been opened, and so nicely excavated that no portion of the entrails, or oily fluids, which would have inevitably destroyed the true colour, was allowed to remain. The natural colour is therefore preserved, which is not of a pale brown, as is commonly imagined, but of a delicate, lovely green ; a colour dependant, it appears, upon a thin internal coating immediately beneath the outer skin, the latter of which is perfectly transparent, and destitute of any colour.

"The pupa of this curious species is represented, together with the perfect insect, on the Vinca rosea.

"There is also a much smaller pupa depicted in the upper part of the plate, that was discovered in one of the islands of the Indian seas, and belonged to the celebrated Mr. Bailey, the astronomer who sailed in one of the expeditions with Captain

ORTHOPTERA.

Cook. This is of an analogous kind to that of the Mantis siccifolia, though evidently distinct. The perfect insect, and in consequence the species, is unknown to us. Our only motive for inserting it is to shew the peculiar singularity of the abdomen, in the middle of which there are two remarkable subquadrangular spots of a filmy texture, that are transparent, and may be seen through very distinctly."—*Donovan*.

Mr. G. R. Gray, in his *Synopsis of the Phasmidæ*, has characterized the last named insect as a distinct species, with the description:—

P. Donnovani (Pupa) : viride, abdomine medio maculis duabus hyalinis subquadratis. Long. Corp. 1—5. Exp. Corp. 6¾.

The chief character of this supposed insect is, I believe, indicative only of the males, notwithstanding the shortness of the antennæ in the accompanying figure, which it is not improbable are fictitious, the antennæ of the male pupæ being much longer, and more numerously articulated, than in the females. If this supposition should be ascertained to be correct, I have little doubt that the figure in question represents merely the male pupa of Ph. siccifolium, and not of a distinct insect. In the Crochard edition of the *Regne Animal*, a figure is given of the male pupa, which is described as the perfect state of that sex; but unfortunately only the three basal segments of the abdomen are represented, the fourth, which bears the ocelli, and the remaining segments being omitted.

LOCUSTA (MONACHIDIA) RETICULATA.

Plate XII. fig. 1.

SECTION. SALTATORIA, *Latreille*.
FAMILY. LOCUSTIDÆ, *Leach*.
GENUS. LOCUSTA.
SUBGENUS. MONACHIDIA. (Monachidium, *Serville*.)
SPECIES. LOCUSTA (MONACHIDIA) RETICULATA : ferruginea ; thorace crista trifida carinata, postice producta cymbiformi, carina lineaque laterali nigris ; tegminibus nigris flavo reticulatis. Long. (alis clausis) unc. 2½.
LOCUSTA (MONACHIDIA) : ferrugineous ; thorax with a three-cleft, keeled, boat-shaped crest produced over the elytra, the ridge and a lateral line of black ; tegmina black, reticulated with yellow. Length with the wings shut, 2½ inches.
SYN. Gryllus reticulatus, *Fabr. Ent. Syst.* 11. p. 47. *Donov. 1st edit.*

Fabricius describes this insect from the Banksian collection, and as an inhabitant of Tranquebar.

ORTHOPTERA.

LOCUSTA (PHYMATEA) PUNCTATA.

Plate XII. fig. 2.

SUBGENUS. PHYMATEA. (Phymateus, *Thunberg, Serville.*)

SPECIES. LOCUSTA (PHYMATEA) PUNCTATA: thorace verrucoso atro, lateribus capitis et thoracis flavis; tegminibus atris, flavo-punctatis; alis atris; abdomine rufo-annulato. Long. (alis clausis) 2¾ unc.

LOCUSTA (PHYMATEA): with the thorax warty, black, with the sides of it and of the head yellow; tegmina black, spotted with yellow; wings black; abdomen with red rings. Length, (with the wings shut,) 2¾ inches.

SYN. Gryllus punctatus, *Drury, App. vol.* 2, *pl.* 41, *fig.* 4. *Fabr. Ent. Syst.* 2. 51. *Thunberg, Mem. Acad. Petersb. p.* 258. *Stoll Saut. pl.* 7. *b*, *f.* 24 *A.*

HABITAT. East Indies.

SCHIZODACTYLA MONSTROSA.

Plate XII. fig. 3.

FAMILY. ACHETIDÆ, *Leach.*

GENUS. SCHIZODACTYLA. Schizodactylus, *Brulle.* Acheta, *Fabricius.* Gryllus, *Latreille.*

SPECIES. SCHIZODACTYLA MONSTROSA: tegminibus alisque caudatis convolutis; corpore luteo-fusco punctis nigris. Long. Corp. 1⅝ unc.

SCHIZODACTYLA: with the tegmina and wings very long, forming a rolled-up tail; body luteo-fuscous with black markings. Length 1¾ inches.

SYN. Gryllus monstrosus, *Drury App. vol.* 2, *pl.* 43, *fig.* 1. *Fabricius Ent. Syst.* 2. 29. (Acheta m.) *Aud. et Brulle Hist. Nat. Ins. t.* 9, *p.* 161. *British Cycl. Nat. Hist. pl. of Orth. Ins. Le Regne Animal (Edit. Crochard,) Ins. pl.* 82, *f.* 1.

Donovan states that "this very singular creature is found in the vicinity of Bengal, where, according to the information of Mr. Fichtel, it is by no means common. It lives under ground, like the Gryllus Campestris, and some other well known analogous European insects of this tribe."

This fact is of considerable interest, since it clearly proves the strong relation which exists between this interesting species and the family of the crickets, Achetidæ; a relation strengthened, moreover, by the general structure of the body; the anal appendages, (one of which is shown in the accompanying figure); the horizontal

ORTHOPTERA.

position of the dorsal part, and the vertical position of the lateral part of the tegmina; as well as the general colour of the insect.

M. Brullé has very judiciously formed this insect into a new group, which he has named Schizodactylus, from the remarkable structure of the tarsi, which are furnished with curious dilated plates, and are four-jointed. It is chiefly on this account M. Brullé has removed this insect to the family of the grasshoppers, with long antennæ, (Gryllidæ, *Leach.* Locustaires, *Latr.*). M. Brullé has also been induced to adopt this arrangement, by the greater length of the maxillary palpi, and the large size of the wings. The last mentioned character is, however, especially that of many of the crickets, (Achetidæ,) for instance, the common house cricket; and the size of the palpi is scarcely a character of first-rate importance. M. Brullé has moreover omitted to notice, that the anterior tibiæ are destitute of that peculiar oval cavity which he expressly states (p. 125) to be found " dans *tous* les insects de cette famille." Hence, from the majority of its characters, I am still induced to place this insect in the family Achetidæ, regarding it as the connecting link between that family and the Gryllidæ. The genus Anostostoma of Gray, (*Mag. Nat. Hist.* 2nd *series,* 1 *p.* 143.) is another very interesting link.

PHYLLOPHORA AMBOINENSIS.

Plate XIII. fig. 1.

FAMILY. GRYLLIDÆ, *Leach.* Locustaires, *Latreille.*

GENUS. PHYLLOPHORA, *Thunberg.* Steirodon, *Serville.* Locusta p. *Fabricius.*

SPECIES. PHYLLOPHORA AMBOINENSIS: " viridissima; thorace tetragona, angulis dentatis; scutello [sc. area interna tegminum] magno fusco-viridi; elytris foliaceis latissimis."—*Donovan.* Long. alis clausis, unc. 4½.

PHYLLOPHORA: green, thorax quadrangular, with the angles dentated; scutel large, brown-green; wing-cases leaf-formed and very broad. Length, with the wings closed, 4½ inches.

SYN. Locusta Amboinensis. *Donov.* 1*st edit.*

" The only specimen we have ever seen of this elegant species, is that delineated in the annexed plate with Locusta citrifolia. This insect was received from Amboyna some years ago by Governor Holford, then resident in India, and is at this time in the collection of the author.

ORTHOPTERA.

"Both the elytra, or wing-cases, and the posterior part of the thorax, are of a fine delicate green; the anterior part of the thorax yellowish brown; with the head and body still paler. The wing-cases, as usual in this tribe, bear no very distant resemblance to the leaves of certain plants, not only in colour, but also in the outline, and still more so in the conformation of the nerves which arise and branch off towards the extremities, exactly in the same manner as the nerves arise, and ramify, from the mid-rib in the leaves of the far greater number of plants. One peculiarity in the structure of the elytra in our new species deserves remark: the scutel, or rather that portion of the wing-case on the left side that folds over the back when the creature is at rest, is of a much stronger texture than any other part of the insect, except the thorax, and serves as an external covering or defence to the corresponding lobe of the other wing-case, which is of a more delicate nature, consisting only of a thin and pellucid membrane, the surface of which is hyaline or glassy. The wings are remarkably tender, of a whitish colour, and semitransparent."—*Donovan.*

PHYLLOPHORA CITRIFOLIA.

Plate XIII. fig. 2.

SPECIES. PHYLLOPHORA CITRIFOLIA: thorace tetragono, angulis crenatis; tegminibus viridibus, folium lauri referentibus. Long. alis clausis, unc. 3.
PHYLLOPHORA: with the thorax quadrangular, crenated at the angles; wing-covers green like a laurel leaf. Length, with the wings closed, 3 inches.

SYN. Gryllus citrifolius, *Linnæus, Mus. Reg. Ulr.* p. 125. *Syst. Nat.* II. 695. *Fabricius, Ent. Syst.* t. 2. p. 33, (Locusta c.) *De Geer Mem. vol.* 3, *pl.* 33, *f.* 3. *Serville, Rev. Orth.* p. 44. (Steirodon c.)

HABITAT. "In Indiis," (*Linnæus.*) Surinam, (*De Geer.*) Cayenne, (*Serville.*)

Donovan states that he received it with the preceding insect from Amboyna.

Order. HEMIPTERA, *Linnæus* (continued).

RAPHIGASTER INCARNATUS.

Plate XIV. fig. 1.

SECTION. HETEROPTERA, *Latreille.*
FAMILY. PENTATOMIDÆ, *Leach.* Cimex, *Linn.*
GENUS. RAPHIGASTER, *Laporte.*
SPECIES. RAPHIGASTER INCARNATUS : supra sanguineus; capite, scutelli maculis duabus; hemelytris unica, membranaque apicali nigris. Long. Corp. unc. 1.
RAPHIGASTER : above sanguineous, with the head; two spots on the scutellum; a single one on each side of the hemelytra, and the apical membrane black. Length 1 inch.
SYN. Cimex incarnatus *Drury Illustrations, App. vol.* 2, 1st edit. *pl.* 36, *f.* 5.
Cimex nigripes, *Fabr. Syst. Rhyng.* 149. 17. *Wolff,* 1, *t.* 2, *f.* 11. *Donovan* 1st edit.
Cimex melanopus, *Gmel. Syst. Nat.* 2149.

This beautiful species bears a distant similitude to Cimex aurantius, described among the insects of China; it is found in the environs of Batavia, as we are informed, but is by no means common there.

TESSERATOMA PAPILLOSA.

Plate XIV. fig. 2.

GENUS. TESSERATOMA, *St. Farg, et Serv. Laporte. Burmeister.*
SPECIES. TESSERATOMA PAPILLOSA : lutea; thoracis lateribus subrotundatis; antennis fuscis basi subferrugineis; abdomine supra purpureo-ferruginono. Long. Corp. 1 unc.
TESSERATOMA; luteous; thorax with the sides rounded; antennæ dark brown, reddish towards the base; abdomen above purplish-ferruginous. Length 1 inch.
SYN. Cimex papillosus, *Drury, App. vol.* 2, *pl.* 43, *f.* 2. (nec *Fabr. Burmeister, &c. Donovan,* 1st edit.
Tesseratoma Sonneratii, *St. Farg. et Serv. Enc. Méth.* 10. 590.
Cimex Chinensis, *Thunberg. Nov. Ins. t.* 11, *f.* 59. *Laporte, Hemipt. p.* 60.

"Professor Thunberg describes this Cimex under the name of Chinensis. Our figures represent both the larva and the perfect insect. That our specimens are from

HEMIPTERA.

the East Indies need not admit of doubt: the same kind is mentioned as a native of Sierra Leone by Fabricius, perhaps without sufficient authority."—*Donovan.*

There is, perhaps, some doubt, as to this species being the C. Chinensis of Thunberg, or the real papillosus of Drury. The African species, described by Fabricius under the latter name, is totally distinct. See my observations on this insect, in the second edition of *Drury's Illustrations, vol.* 2, *p.* 91.

PENTATOMA CRUCIATA.

Plate XIV. fig. 3.

GENUS. PENTATOMA, *Latreille.* Cimex, *Fabricius.* Burm.
SPECIES. PENTATOMA CRUCIATA: nigro pallidoque varia; scutello nigro, cruce alba. Long. Corp. unc. $\frac{1}{3}$.
PENTATOMA: pale yellow with black spots, (10 on the thorax, 4 in front, and 6 behind); scutellum black, with a white cross. Length $\frac{1}{3}$ of an inch.
SYN. Cimex cruciata, *Fabr. Ent. Syst.* 4. 119. *Syst. Rhyng. p.* 174. *Wolff Cimic. t.* 7, *f.* 59.

Fabricius states that this species was found in the East Indies by Dr. Kœnig. Donovan's specimen was from Bengal.

PENTATOMA MACTANS.

Plate XIV. fig. 4.

SPECIES. PENTATOMA MACTANS: ovata depressa; supra rufa, thorace scutelloque maculis duabus; hemelytrorumque membrana apicali nigris. Long. Corp. $\frac{1}{2}$ unc.
PENTATOMA: ovate depressed; above rufous, with two black spots on the thorax, and two on the scutellum; apical membrane of the hemelytra black. Length $\frac{1}{2}$ inch.
SYN. Lygæus Mactans, *Fabr. Ent. Syst.* 4. 161. *Syst. Rhyng. p.* 227.
Cimex Mactans, *Fabr. Sp. Ins.* 2. 366. *Donovan,* 1st *edit.*

ACANTHOSOMA UNIGUTTATA.

Plate XIV. fig. 5.

GENUS. ACANTHOSOMA, *Curtis.* Cimex, *Donovan.*
SPECIES. ACANTHOSOMA UNIGUTTATA: thorace acute spinoso; ferruginea, scutello puncto magno albo notato. Long. Corp. unc. $\frac{1}{2}$.

HEMIPTERA.

ACANTHOSOMA : with the thorax produced on each side into an acute spine; ferruginous, with a large white spot at the base of the scutellum. Length $\frac{1}{2}$ an inch.

SYN. Cimex uniguttatus, *Donovan, 1st edit.*

Donovan states that this curious species is a native of Madras.

RHYNCHOCORIS VIRIDIS.

Plate XIV. fig. 6.

GENUS. RHYNCHOCORIS, *Westwood.* Cimex, *Donovan.*

SPECIES. RHYNCHOCORIS VIRIDIS : thorace spinoso ; viridis ; hemelytris fusco-cinereis ; scutello apice flavescenti ; thoracis spinis cylindricis truncatis. Long. Corp. unc. *.

RHYNCHOCORIS : with the thorax produced behind on each side into a cylindric spine, truncate at the tip ; green ; hemelytra brownish-grey ; scutel at the tip yellowish. Length $\frac{4}{8}$th of an inch.

SYN. Cimex viridis, *Donovan, 1st edit.*

This rare insect is from Ceylon.

RHYNCHOCORIS HAMATA.

Plate XIV. fig. 7.

SPECIES. RHYNCHOCORIS HAMATA : thorace acute spinoso, testacea ; elytris virescentibus ; abdomine serrato ; denticulis nigris. Long. $\frac{9}{10}$ unc.

RHYNCHOCORIS : with the thorax produced into an acute spine on each side, testaceous buff, with greenish elytra ; the margin of the abdomen serrated ; the teeth black. Length $\frac{9}{10}$th of an inch.

SYN. Edessa hamata, *Fabr. Ent. Syst.* 4. 104. *Syst. Rh.* p. 147.
Cimex serratus, *Donovan, 1st edit.*
Rhynchocoris hamata, *Westwood, in Cat. Hemipt. Mus. Hope.*

Java, and other parts of the East Indies.

Order. **LEPIDOPTERA**, *Linnæus.*

PAPILIO ANTENOR.
Plate XV. fig. 1.

TRIBE.   DIURNA, *Latreille.*   GENUS: Papilio, *Linn.*, (Butterflies.)
FAMILY.   PAPILIONIDÆ, *Leach.*
GENUS.   PAPILIO, *Latreille.* (Papiliones Equites, *Linn.*)
SPECIES.   PAPILIO ANTENOR : alis dentatis concoloribus atris albo-maculatis ; posticis caudatis ; disco atomis viridibus lunulisque marginalibus rubris. Expans. alar. 6 unc. 6 lin.
         PAPILIO : with the wings dentate, black with many white spots, the posterior pair tailed, and with the disk powdered with green atoms, the margin with red lunules. Expanse of the wings 6½ inches.
SYN.   Papilio Antenor, *Drury, Ill. Exot. Ent.* vol. 2 app. pl. 3, *f.* 1. 2d edit. v. 2, *p.* 4. *Fabr. Ent. Syst.* III. I. *p.* 4, *No.* 9. *Jablonsky, Pap.* 2. *t.* 3, *fig.* 1. *Boisduval Hist. Nat. Lepid.* 1, *p.* 189, *No.* 2. *Enc. Méth.* 9, 69.
HABITAT.   Tropical Africa.

This species was originally described by Drury, from whom all subsequent writers (including Donovan) have derived their descriptions and figures of this exceedingly rare butterfly. Drury states that he was unaware from what country his specimen was obtained. Fabricius, however, gave India as its country, on which authority, apparently, Donovan introduced it into this work, observing that it might " be numbered with much propriety among the rarest of the Papilio tribe found in India." In the second edition of Drury's Illustrations, I have stated the incorrectness of this locality, on the authority of a specimen in the collection of the Rev. F. W. Hope, brought from tropical Africa by the late Mr. Ritchie. Had the insect moreover been an inhabitant of India, it would surely have been received more frequently. No specimen however, as far as I am aware, is contained in any of the continental collections.

PAPILIO ANTIPHUS.
Plate XV. fig. 2.

SPECIES.   PAPILIO ANTIPHUS : alis concoloribus nigris, anticis ad apicem griseo radiatis, posticis caudatis, lunulis septem submarginalibus rubris. Expans. alar. 3 unc. 3 lin.
         PAPILIO : with all the wings similarly coloured, black ; the anterior paler, greyish

LEPIDOPTERA.

at the tips, with black rays; the posterior pair tailed, and with seven red lunules. Expanse of the wings 3¼ inches.

SYN. Papilio Antiphus, *Fabr. Ent. Syst.* III. 1. p. 10. *Enc. Méth.* 9. 71. *Boisduval Hist. Nat. Lepid.* vol. 1, p. 266.

VAR. Papilio Polygius, *Enc. Méth.* 9. 811.

HABITAT. India, (*Fabricius*). Philippine Islands, (*Enc. Méth.*) Manilla, (*Boisduval*).

Nearly allied to Papilio Polydorus, but one-fifth smaller, and without the white spot on the posterior wings. It was originally described by Fabricius from the unpublished collection of drawings of Mr. Jones, from which it is most probable that Donovan, who had access thereto, obtained this figure, which is the only one yet published of the species. In the *Encyclopédie Méthodique*, a variety of this species, with nine red lunules, is described under the name of Papilio Polygius, from the Philippine Islands, where this insect appears to replace Papilio Polydorus.

ORNITHOPTERUS PRIAMUS.

♂ Plate XVI.

GENUS. ORNITHOPTERUS. (Ornithoptera, *Boisduval.*)

SPECIES. ORNITHOPTERUS PRIAMUS : alis holosericeis supra viridibus, limbo nigro ; anticis fascia longitudinali latissima ; posticis maculis submarginatibus, nigris ; his denticulatis abdomine flavo ♂. Expans. alar. unc. 8½.

ORNITHORPTERUS : with the wings silky ; above green with a black border, the anterior with a very broad longitudinal black bar, the posterior with black submarginal spots, the latter wings denticulated, the abdomen yellow, ♂. Expanse of the wings 8½ inches.

SYN. Papilio (Equ. Tr.) Priamus, *Linn. Amœn. Acad.* 5. 3. f. 203. *Syst. Nat.* 2. 744. *Clerck Icones*, t. 17. *Fabr. Ent. Syst.* III. 1. p. 11. *Cramer, pl.* 23, *A. B. Enc. Méth.* 9, p. 25. *Boisduval Hist. Nat. Lepid.* vol. 1, p. 173.

Donovan observes, that " With the exception of Papilio Ulysses, which perhaps in point of splendour may excel, Papilio Priamus is beyond comparison the most lovely creature of this tribe of insects hitherto discovered, either in India or any other country. It is a native of Amboyna, where we understand it is extremely rare, and bears a considerable price among the Dutch amateurs in that island. We obtained a pair of them in fine condition some years ago from the cabinet of the late Mr. Tunstall, who had purchased them in Holland, from a collection made by one of the Dutch governors in Amboyna. This rarity is figured in a resting position on the blossoms of the Mimosa Grandiflora."

LEPIDOPTERA.

PAPILIO EMPEDOCLES.

Plate XVII. fig. 1.

Species. Papilio Empedocles : alis fuscis ; anticis utrinque fascia maculari viridi, transversa media, posticis caudatis, subtus ad angulum ani lunulis duabus nigris. Expans. alar. unc. $4\frac{1}{2}$.

Papilio : with the wings brown ; the anterior on both sides with a transverse central band of green spots ; the posterior tailed and with two black lunules at the anal angle beneath. Expanse of the wings $4\frac{1}{2}$ inches.

Syn. Papilio (Nymphalis) Empedocles. *Fab. Ent. Syst.* III. 1. p. 70. *Enc. Méth.* 9. 810. *Boisduval Hist. Nat. Lepid.* vol. 1, p. 29.

Habitat. " India Orientali," (*Fabricius.*) Island of Bourbon, (*Boisduval.*)

The original Fabrician description was taken from a specimen in the collection of Sir Joseph Banks, now in the possession of the Linnæan Society. By Fabricius this species was inserted, by some strange oversight, amongst the Nymphales.

PAPILIO DEIPHOBUS.

♂ Plate XVII. fig. 2.

Species. Papilio Deiphobus : alis nigris subtus basi rubro-maculatis ; posticis caudatis maculis septem rubris subannularibus submarginalibus. ♂. Expans alar. unc. 6.

Papilio : with black wings, the base beneath being spotted with red ; the posterior pair tailed, with seven red subannular and submarginal spots. Expanse of the wings 6 inches.

Syn. Papilio (Eq. Tr.) Deiphobus, *Linn. Syst. Nat.* 2. 746. *Clerck Icon. t.* 25. *Cramer, t.* 181, *A.B. Fabricius Ent. Syst.* III. 1. p. 5. *Enc. Méth.* 9. 64. *Boisduv. Hist. Nat. Lepid.* vol. 1, p. 200.

♀ Papilio Alcandor, *Cramer, pl.* 40, *A. B.*

The female has the wings brown above, the posterior pair being ornamented with a large palmated buff mark in the centre. Boisduval gives the Moluccas, Amboyna, and Celebes, as the habitat of this species.

LEPIDOPTERA.

PAPILIO LACEDEMON.

Plate XVII. fig. 3.

SPECIES.   PAPILIO LACEDEMON : alis dentatis nigro-fuscis, lunulis marginalibus albis ; posticis ecaudatis subtus brunneis lunulis submarginalibus nigris.   Expans. alar. unc. $4\frac{1}{4}$.

PAPILIO : with the wings dentated, brownish-black, with marginal lunules ; the posterior without tails, beneath brown, with black submarginal lunules. Expanse of the wings $4\frac{1}{4}$ inches.

SYN.   Papilio (Eq. Ach.) Lacedemon, *Fabr. Ent. Syst.* III. 1, p. 36.   *Boisduval Hist. Nat. Lepid.* vol. 1, p. 374.   *Enc. Méth.* 9. 38.

HABITAT.   Malabar, (*Fabricius*).

ORNITHOPTERUS REMUS.

Plate XVIII.

SPECIES.   ORNITHOPTERUS REMUS : alis nigris ; posticis dentatis, subtus albis, his utrinque maculis submarginalibus auratis.   Expans. alar. unc. $9\frac{1}{2}$.

ORNITHOPTERUS : with black wings ; the posterior pair dentate, beneath (except at the base) white, with golden coloured submarginal spots more or less expanded. Expanse of the wings $9\frac{1}{2}$ inches.

SYN.   Ornithoptera Remus, *Boisduval Hist. Nat. Lepid.* p. 176, (nec. Priamus ♀ p. 173).
Papilio Remus, *Fabr. Ent. Syst.* III. 1, p. 11.   *Cramer, pl.* 135 *A*, 136 *A*, and 386 *A. B.   Enc. Méth.* 9, p 126.   *Naturalist's Lib. Entomol.* v. 5, pl. 1, f. 2.
Papilio Panthous, *Donovan, 1st edit.*

Donovan observes of this species, that " it is one of the many magnificent natural productions of Amboyna, and is extremely rare." By Donovan it was given as the Papilio Panthous, with the further observation, " The supposed female of this insect is also considered as the female of Papilio Priamus ; it is a gigantic creature, being still larger than the butterfly represented, but it is less beautiful, and the general colour an obscure reddish brown." The insect here referred to by Donovan is the real Papilio Panthous, and is now generally considered to be the true female of P. Priamus. The insect here figured is evidently a male, as may be seen from the anal valve, which so peculiarly distinguishes the males of the Ornithopteri. Hence Godart and Boisduval must be in error in giving this figure as the female of Priamus. In the nearly uniform colour of the upper wings it better agrees with P. Remus, although the markings of the posterior wings are different. I have not, however,

LEPIDOPTERA.

ventured to give it as a distinct species, considering it possible that the markings may be liable to variation. It is figured resting on a flower of the double variegated Japan rose, a scarce variety of the Camellia Japonica.

In the Naturalist's Library, Entomology, vol. 5, p. 89, this figure is given as the female of Priamus, although a figure agreeing with the present is likewise given under the name of Remus.

ORNITHOPTERUS HELIACON.

Plate XIX. fig. 1.

SPECIES. ORNITHOPTERUS HELIACON: alis dentatis anticis nigris, posticis flavis limbo tenui undata punctisque submarginalibus nigris. Expans. alar. unc. 5½.

ORNITHOPTERUS: with the wings dentated; the anterior black, the posterior golden yellow, with a festooned black margin, and with a row of black submarginal spots, of which the two exterior are the largest, the others being often obsolete. Expanse of the wings 5½ inches.

SYN. Papilio (Eq. Tr.) Heliacon, *Fabr. Ent. Syst.* III. 1. *p.* 19. 60. *Boisduval Hist. Nat. Lepid. p.* 178. (Ornithoptera H.)
Papilio Amphrisius, ♂ *Enc. Méth.* 9. 27. *Horsfield Lep. Jav. pl.* 4. *f.* 3.
Papilio Astinous, *Fab. Ent. Syst.* III. 1. *p.* 19.
Papilio Minos, *Cramer, pl.* 195 *A.*
Papilio Pompeus, *Cramer, pl.* 25 *A.*
Amphrisius Nymphalides, *Swainson Zool. Illustr. new series, pl.* 98.

Originally described by Fabricius from the Banksian Collection and Mr. Jones's drawings; and as the present figure exactly agrees with the Fabrician description, it is most probable that it was from one of these sources (to both of which Donovan had access) that this figure was taken. This is of some importance, as the species has been confounded with several others, as may be seen from the synonyms quoted above, and as it is liable to considerable variation in the spots of the posterior wings, and in the occasional paler radiation of the superior wings. In a specimen which I possess, exactly corresponding with this figure, the neck has a scarlet band, and there are some scarlet bands at the sides of the thorax beneath.

The transformations of this species have been observed by Dr. Horsfield in Java, and figures of the larva and pupa published in the *Lepidoptera Javanica* (*pl.* 4, *f.* 13). The caterpillar is thick, of a yellowish colour, with a broad dorsal white line. It is armed with eight rows of erect obtuse fleshy appendages, as well as with a furcate

LEPIDOPTERA.

tubercle behind the head. The chrysalis is thick, curved and armed on the back of each of the three intermediate abdominal segments with a pair of conical points.

Dr. Horsfield considers this insect as the type of the last of the three great sections of the genus Papilio having the antennæ with obscure annuli, (vide synopt. table). Mr. Swainson gave it as a subgenus named Amphrisius, confounding the two species Heliacon and Amphrisius together, and applying the latter name as that of the genus. But as Priamus and its allies are the *types* of the same group, it would be absurd, even were it not inconvenient, to retain this genera name. I have therefore employed M. Boisduval's name of Ornithoptera, altering its termination, however, to make it agree with the sex of the old genus Papilio, from which it has been separated.

PAPILIO IDÆUS.

Plate XIX. fig. 2.

SPECIES : PAPILIO IDÆUS : alis nigris, anticis fascia abbreviata flava, posticis dentatis, macula palmata trifida punctisque sanguineis, emarginaturis albidis. Expans. alar. unc. 4¼.

PAPILIO : with black wings, the anterior with an abbreviated yellow band, the posterior dentated, with a palmated mark composed of three red united spots, and several other red spots, the emarginations yellowish. Expanse of the wings 4¼ inches.

SYN. *Fabricius, Ent. Syst.* III. 1. *p.* 16. *Enc. Méth.* 9. 33. *Boisduval Hist. Nat. Lepid. p.* 299.

HABITAT. Madras, *Fabricius.*

This species was originally described from the collection of Mr. Drury and Mr. Jones's drawings, which evidently supplied Mr. Donovan with the present figure. The French Entomologists are only acquainted with the species through the works of Fabricius and Donovan. M. Boisduval is of opinion, that from the facies of this species it is rather an inhabitant of South America than the East Indies.

PAPILIO ASTYANAX.

Plate XX. fig. 1.

SPECIES. PAPILIO ASTYANAX : alis nigris concoloribus, anticis fascia sesquialtera alba striata, posticis dentatis, ecaudatis, macula discoidali palmata lunulisque sex submarginalibus sanguineis. Expans. alar. unc. 4.

PAPILIO : with all the wings black, the anterior having a broad oblique whitish band, striped with black beyond the middle, united to another shorter, which

LEPIDOPTERA.

crosses the discoidal cell; the posterior wings dentated without tails, with a discoidal palmated blotch, and six submarginal lunules of a sanguineous colour. Expanse of the wings, 4 inches.

SYN. Papilio Astyanax, *Fabricius, Ent. Syst.* III. 1. *p.* 13. *Enc. Méth.* 9. 72. *Boisduval, Hist. Nat. Lepidopt.* 271.
Papilio Mutius, ♀ ? *Fabr. loc. at p.* 3.

It is most probable that this figure represents an injured individual of Papilio Mutius, which differs only from Astyanax in possessing tails to the posterior wings. This might have occurred through the tails having been broken off, or as in Papilio Polydorus, some individuals of P. Mutius may possess only very rudimental tails. The species was originally described by Fabricius from the collection of Mr. Drury, and the drawings of Mr. Jones, from which it is probable Mr. Donovan obtained this figure, and from which alone the French Entomologists are acquainted with the species. I possess the late Mr. Haworth's specimens of P. Mutius, which together with the remainder of that gentleman's collection of Papilionidæ, were compared with Mr. Jones's drawings, and which are labelled "P. Astyanax?" the mark of doubt being probably added from the difference in the tail.

The P. Astyanax of the earlier works of Fabricius, is a totally different insect, and a native of America.

PAPILIO POLYMNESTOR.

Plate XX. fig. 2.

SPECIES. PAPILIO POLYMNESTOR: alis dentatis, nigris; anticis fascia lata transversa submarginali albida, (venis interrupta); posticis apice cærulescenti-plumbeis, maculis atris ordinatim digestis. Expans. alar. unc. $6\frac{1}{2}$.

PAPILIO: with the wings dentated, black; the anterior having a broad transverse submarginal whitish band, (interrupted by the veins); the posterior wings with the terminal half-leaden-blue, with two rows of black spots. Expanse of the wings, $6\frac{1}{2}$ inches.

SYN. Papilio (Eq. Tr.) Polymnestor, *Fab. Ent. Syst.* III. 1. *p.* 18. *Cramer, tab.* 53, *figs. A, B.*
Papilio Polymnestor, *Enc. Méth.* 9. 29. *Boisduval, Hist. Nat. Lepid. p.* 191.

HABITAT. "Various parts of India," (*Donovan*). Cachemere, Ceylon, (*Boisduval*).

LEPIDOPTERA.

PAPILIO ULYSSES.

Plate XXI.

SPECIES. PAPILIO ULYSSES: alis nigris, disco cæruleo radiante, posticis caudatis; subtus ocellis septem. Expans. alar. unc. 5¼.

PAPILIO: with black wings, the disc being blue and radiated, posterior wings tailed; beneath brown, with seven large submarginal ocelli on the posterior wings. Expanse of the wings, 5¼ inches.

SYN. Papilio (Eq. Ach.) Ulysses, *Linn. Syst. Nat.* 2. 748. *Fabr. Ent. Syst.* III. 1. p. 23. *Enc. Méth.* 9. 65. *Cramer, tab.* 121, *f. A, B. Boisd. Hist. Nat. Lepid.* 1, p. 202.

♀ Papilio Diomedes, *Linn. Syst. Nat.* 2. 749. *Cramer, t.* 122, *fig. A.*

HABITAT. "The Dutch spice islands," (*Donovan*). Amboyna, Celebes, (*Boisduval*).

CASTNIA EVALTHE.

Plate XXII.

SECTION. ———? (Hesperi sphinges, *Latreille.*)

FAMILY. CASTNDIIÆ, *Westw.*

GENUS. CASTNIA, *Latr. God. Dalman.* Papilio, (Festivi,) *Fabr.*

SPECIES. CASTNIA EVALTHE: alis integris nigris, nitidis; anticis utrinque fasciis duabus, posticis unica flavis; his supra serie e maculis submarginalibus, subtus pagina omni rubris; fascia maculari flava. Expans. alar. 4¼ unc.

CASTNIA: with the wings entire black, shining; the anterior above and beneath with two bands, and the posterior with a single (sometimes interrupted) band of a yellow colour, the latter have also a submarginal row of red spots, the under surface is also red, with a central row of buff spots. Expanse of the wings, 4¼ inches.

SYN. Papilio (Festiv.) Evalthe, *Fab. Ent. Syst.* III. 1. p. 45. *Syst. Ent.* p. 480. *Herbst. Pap. t.* 137, *f.* 1, 2. *Enc. Méth.* 9. 797. (Castnia E.) *Donovan, 1st edition.* (Papilio E.)

Papilio Dardanus, *Cramer, pl.* 17, *fig. E, F.*

HABITAT. "In Indiis," (*Fabricius*).* Surinam and Brazil, (*Enc. Méth.*)

This scarce and beautiful species, represented on a sprig of the Vitex Negundo, Fine-leaved Chaste tree, is here misplaced, not only on account of its belonging to

---

* By the term "In Indiis," Fabricius meant either the West or the East Indies; although he more generally employed the term "In India," or "In India orientali," to designate the habits of East Indian species.

LEPIDOPTERA.

the section of the Hesperi-sphinges of Latreille, but also, because it and all the other species of the remarkable genus to which it belongs, are inhabitants of the tropical parts of South America.

VANESSA LETHE.

Plate XXIII. fig. 1.

FAMILY. NYMPHALIDÆ, *Leach*.
GENUS. VANESSA, *Latr. God. Curtis.* Hamadryades, *Hubn.* Papilio (Nymph. Phal. *Fabr.*)
SPECIES. VANESSA LETHE : alis subcaudatis supra fulvis ; anticis apice late nigro, fascia maculisque flavis ; posticis margine flavo striga nigra. Expans. alar. $2-2\frac{1}{2}$ unc.
VANESSA: with the posterior wings shortly tailed, fulvous above ; the anterior wings with the extremity broadly black, in which is an oblique fascia and several spots of a yellowish colour; the posterior wings have the margin yellowish, with several parallel narrow black submarginal lines. Expanse of the wings, $2-2\frac{1}{2}$ inches.
SYN. Papilio (Nymph.) Lethe, *Fabr. Ent. Syst.* III. 1, *p.* 80. *Enc. Méth.* 9. 818. (Vanessa L.)
HABITAT. " In Indiis," (*Fabricius*). Brazil, (*Enc. Méth.*)

Like the last species, the term applied by Fabricius to indicate the habitat of this species, evidently led Donovan to give it as an inhabitant of the East Indies. It is, however, according to the *Encyclopédie Méthodique*, as well as its very near ally, Vanessa Zabulina, an inhabitant of Brazil.

NYMPHALIS COCLES.

Plate XXIII. fig. 2.

GENUS. NYMPHALIS, *Latreille.* Papilio (Nymphalis) *Fabricius.* Paphia, *Horsfield.*
SPECIES. NYMPHALIS COCLES : alis albo, fusco, flavescentique strigosis, fascia media communi alba; posticis subcaudatis; subtus albidis striga punctorum ocellatorum. Expans. alar. unc. 2.
NYMPHALIS : with the wings transversely streaked with white, brown and yellowish, and with a narrow white bar running across all the wings beyond the middle ; posterior wings with short tails; beneath whitish with a row of eyelets. Expanse of the wings, 2 inches.

LEPIDOPTERA.

Syn. Papilio (N.) Cocles, *Fabr. Ent. Syst.* III. 1, *p.* 65. *Enc. Méth.* 9. 362, (Nymphalis C.)
Habitat. Siam, (*Mus. Banks.*)

This insect, together with P. Periander, figured in plate 37, and some other allied species, constitute a very well marked group in the great genus Nymphalis.

NYMPHALIS (CHARAXES) TIRIDATES.

Plate XXIII. fig. 3.

Genus. Nymphalis. Subgenus : (Charaxes, *Boisduval.* Jasia, *Swainson.*)
Species. Nymphalis (Charaxes) Tiridates : alis supra atro-cæruleis ; margine postico lunulis ochraceis ; omnium dimidio apicali punctis cærulescentibus. Expans. alar. unc. 4¼.
Nymphalis : with blue-black wings ; the posterior margin with ochre-coloured lunules ; the apical half of all the wings with cærulescent spots. Expanse of the wings, 4¼ inches.
Syn. Papilio (Nymph.) Tiridates, *Fabr. Sp. Ins.* 2, *p.* 11. *Ent. Syst.* III. 1, *p.* 62. *Cramer, pl.* 162, *f. A, B. Herbst. Pap. t.* 62, *f.* 3, 4. *Enc. Méth.* 9. 354. (Nymphalis T.) *Drury Ill. v.* 3, *pl.* 23, *f.* 1, 2.
Habitat. Amboyna, (*Fabricius*). Amboyna and Java, (*Enc. Méth.*) Brazil, (*Drury*, but incorrectly?).

IDEA AGELIA.

Plate XXIV.

Family. Heliconiidæ, *Swainson.*
Genus. Idea, *Fabricius,* (*Syst. Gloss. in Illig. Mag.*) Papilio, (Dan. Cand.) *Linnæus.*
Species. Idea Agelia : alis ovatis integerrimis albo-denudatis, venis maculisque nigris : anticis margine externo nigro serie punctorum alborum. Expans. alar. unc. 6¼.
Idea : with ovate and very entire wings, semi-transparent white, with black veins and spots ; the outer margin of the anterior wings being black with a row of white spots. Expanse of the wings, 6¼ inches.
Syn. Idea Agelia, *Enc. Méth.* 9. 195. *Audouin et Brullé Hist. Nat. Ins. Lep. t.* 2, *pl.* 5.
Papilio (D. C.) Idea, *Linn. Syst. Nat.* 2. 758. *Fabr. Ent. Syst.* III. 1. *p.* 185. *Clerck Icon. tab.* 38, *f.* 1.

Donovan observes of this insect—" There is a peculiar delicacy in the appearance of this large and rare Papilio that strongly recommends itself to our attention, and,

LEPIDOPTERA.

notwithstanding that the figure of it has a place already in several works on Entomology, impels us to include it in the present selection of Indian insects. Clerk, Cramer, and Drury, have each given a delineation of it, the latter of whom names it Papilio Lynceus; but it is almost superfluous to add, that it is beyond dispute the Papilio Idea of Linnæus. Our specimens are from Amboyna, and are represented on the common Indian Yellow Jasmine."

Although it is evident that this figure of Donovan's represents the true Linnæan Papilio Idea, that figured by Drury is a distinct species, for which the name of Idea lyncea must be retained, Dr. Horsfield has figured another species, from Java, under the name of Idea Gaura, (*Lep. Jav. pl.* 6, *f.* 1. *Boisduval Lepid. pl.* 11, *f.* 11). A fourth, Idea Daos, is figured by M. Boisduval, in the last named work, (*pl.* 24, *fig.* 3); and I believe a fifth was in the collection of General Hardwick, now at the British Museum.

PIERIS HIPPIA.

Plate XXV. fig. 1.

FAMILY. PAPILIONIDÆ, *Leach.*
GENUS. PIERIS, *Schrank, Latr. God.* Papilio (Dan. Cand.) *Linnæus.*
SPECIES. PIERIS HIPPIA : alis repandis virescenti-albis ; venis limboque (albo punctato) nigris.
Expans. alar. unc. 3¼.
PIERIS : with broad greenish white wings ; the veins and borders black, the latter often with white spots. Expanse of the wings, 3¼ inches.
SYN. Papilio (Festiv.) Hippia, *Fabr. Ent. Syst.* III. 1. *p.* 59. *Enc. Méth.* 9. 19. 3, (Danais H.)
Pieris Valeria, *Cramer, pl.* 85, *f.* A. *Enc. Méth.* 9. 155. *Boisduval Hist. Nat. Lep.*
p. 444.
HABITAT. "Poulicandor, *Mus. Banks,*" (*Fabricius*). Java, India, (*Boisduval, Enc. Méth.*)

This species has somewhat the facies of some of the Danai, which probably induced Fabricius to remove it from its legitimate situation.

DANAIS AFFINIS.

Plate XXV. fig. 2.

FAMILY. HELICONIIDÆ, *Swainson.*
GENUS. DANAIS, *Latreille.* Euplœa, *Fabricius.* Papilio (Dan. festiv. p. *Linn.*)
SPECIES. DANAIS AFFINIS : alis nigris albo maculatis ; posticis dentatis, subtus limbo fusco punctis flavis albisque ordinatim digestis. Expans. alar. unc. 3¼.

LEPIDOPTERA.

DANAIS: with the wings black, spotted with white; the posterior dentate, and beneath with a broad brown border, with one row of fulvous, and two rows of smaller white spots. Expanse of the wings, 3¼ inches.

SYN. Papilio (Fest.) affinis, *Fabr. Ent. Syst.* III. 1. p. 58, 181. *Syst. Ent.* p. 511. *Enc. Méth.* 9, 182.

HABITAT. "Nova Hollandia, *Mus. Dom. Banks*," (*Fabricius*).

I know not upon what authority Donovan introduced this species amongst the inhabitants of India, New Holland being the only locality recorded by Fabricius, by whom alone the insect has been described, the French Entomologists merely copying his description.

NYMPHALIS PHILOMELA.

Plate XXV. fig. 3.

SPECIES. NYMPHALIS PHILOMELA: alis nigris, anticis albo punctatis et lineatis; posticis basi flavo-radiatis, apice nigris albo-punctatis. Expans. alar. unc. 3.

NYMPHALIS: with the wings black, the anterior with white spots and lines, the posterior radiated with yellow at the base, with a black border spotted with white. Expanse of the wings, 3 inches.

SYN. Papilio (Fest.) Philomela: *Ent. Syst.* III. 1, p. 57. *Enc. Méth.* 9. 398, (Nymphalis? Ph.)

HABITAT. "In Indiis, *Mus. Dom. Banks*," (*Fabricius*).

PIERIS (IPHIAS) LEUCIPPE.

Plate XXVI. fig. 1.

GENUS. PIERIS, *Schrank, Latreille, Godart.* Colias, *Horsfield.* (SUBGENUS: Iphias, (*Boisduval.*)

SPECIES. PIERIS (IPHIAS) LEUCIPPE: alis anticis supra vivide fulvis, venis, margineque nigris; posticis supra flavis, subtus saturatioribus atomis fuscis. Expans. alar. unc. 4½.

PIERIS (IPHIAS): with the anterior wings above rich orange, with black veins and border; the posterior pair above yellow, beneath darker with brown atoms. Expanse of the wings, 4½ inches.

SYN. Papilio (Dan. Cand.) Leucippe, *Fabr. Sp. Ins.* 2, p. 44. *Ent. Syst.* III. 1, p. 198. *Cramer Pap.* pl. 36, fig. A, C. *Herbst. Pap.* t. 109. *Enc. Méth.* 9. 119. (Pieris L.) *Boisduval Hist. Nat. Lepid.* 1. 596. (Iphias L.)

HABITAT. The Island of Amboyna.

LEPIDOPTERA.

ANTHOCHARIS DANAE.

Plate XXVI. fig. 2.

GENUS. ANTHOCHARIS, *Boisduval.* (Pontia p. *Horsf.* Pieris p. *Enc. Méth.*)

SPECIES. ANTHOCHARIS DANAE: aliis rotundatis albis; anticis apice coccineis, margine fasciaque nigris; singulis subtus striga moniliformi punctorum subferruginea abbreviata. Expans. alar. unc. 2.

ANTHOCHARIS: with the wings rounded and white; the anterior with a large apical spot of red crimson-pink, surrounded with black; beneath all the wings have a short row of dull moniliform spots. Expanse of the wings, 2 inches.

SYN. Papilio (D. C.) Danae, *Fab. Ent. Syst.* III. p. 203. *Syst. Ent. p.* 476. *Enc. Méth.* 9. 124. (Pieris D.) *Boisduval Lepid.* i. 570. (Anthocharis D.) *Horsfield Lep. Jav. p.* 141.

Papilio Eburia, *Cramer Pap. pl.* 352. *C, D, E, F.*

HABITAT. "India orientali," (*Fabricius*). Bengal, and the Cape of Good Hope, (*Boisduval*). Mysore, (*Donovan*).

GONIAPTERYX MÆRULA.

Plate XXVII. fig. 1.

FAMILY. PAPILIONIDÆ, *Leach.*

GENUS. GONIAPTERYX, *Westwood.* Gonepteryx, *Leach, Curt. Steph.* Colias, *Latr.* Rhodocera, *Boisduval Dup.*

SPECIES. GONIAPTERYX MÆRULA: alis angulatis flavis; anticis supra puncto medio atro; singulis subtus disco macula ocellari oblongo. Expans. alar. unc. 3, lin 4.

GONIAPTERYX: with angulated yellow wings; the anterior with a central black spot on the upper side; all the wings beneath with an oblong ocellated discoidal spot. Expanse of the wings, 3½ inches.

SYN. Papilio (Dan.) Mærula, *Fabricius Ent. Syst.* III. 1. p. 212. *Ent. Syst. p.* 479. *Boisduval Hist. Nat. Lepid.* 1, p. 600. (Rhodocera M.) *Enc. Méth.* 9. 89. (Colias M.) *Boisdv. et Leconte, Icon. Lep. Am. Sept. pl.* 23.

Papilio Ecclipsis, *Cramer pl.* 129, *f. A, B.*

Fabricius gave "India" as the habitat of this species, whence Donovan introduced it into this work. It is now, however, well ascertained that it, as well as all the other species of the same genus, (except the British G. Rhamni,) are inhabitants

LEPIDOPTERA.

of the New World. Jamaica, and Florida, are given as its true habitats by Boisduval, and New York is added in the Encyclopédie Méthodique.

M. Boisduval's grounds for substituting his own generic name, Rhodocera, in preference to that of Dr. Leach, are quite untenable; Dr. Leach's name having a long priority over that of Latreille, employed for Noctua, Libatrix, which had also been previously named Scoliopteryx by Germar, which name Stephens has retained. I have been compelled, however, slightly to alter Dr. Leach's name, to render it more in accordance with its Greek derivatives.

PIERIS JUDITH.

Plate XXVII. fig. 2.

SPECIES. PIERIS JUDITH : alis rotundatis integerrimis ; anticis albis venis margineque postico (albo maculato) nigris ; posticis fulvis, margine nigris. Expans. alar. unc. 2—2½.
PIERIS : with the wings entire and rounded ; the anterior white, with the veins and posterior margin, black, the latter spotted with white ; the posterior wings fulvous, with a black border. Expanse of the wings, 2—2½ inches.

SYN. Papilio (Dan. Cand.) Judith, *Fabr. Ent. Syst.* III. 1. *p.* 202. *Boisduval Hist. Nat. Lepid.* 1. 468. *Enc. Méth.* 9. 121. *Hubner Zütt.* 669—670. (Acræa J.) *Horsfield Lep. Jav. p.* 144. (Pontia J.)

HABITAT. " Poulicandor, *Mus. D. Banks.*" (*Fabricius*). Java, Sumatra, (*Boisduval*)

PIERIS LIBYTHEA.

Plate XXVII. fig. 3.

SPECIES. PIERIS LIBYTHEA : alis rotundatis integerrimis albis ; anticis costa baseos apiceque. posticis punctis marginalibus fuscis. Expans. alar. unc. 2⅛.
PIERIS : with the wings entire, rounded, and white ; the anterior having the costa at the base and the extremity, and the posterior having several marginal spots dark brown. Expanse of the wings, 2⅛ inches.

SYN. Papilio (Dan. Cand.) Libythea, *Fabr. Ent. Syst.* III. 1. *p.* 190. *Syst. Ent. p.* 471. Pieris Libitina, *Enc. Méth* 9. 133. *Boisduval Hist. Nat. Lepid. p.* 499.

The Fabrician *specific* name of this insect has been altered in the Encyclopédie Méthodique, merely because the same name has been *subsequently* used *generically* for a distinct group of Lepidoptera.

LEPIDOPTERA.

ANTHOCARIS EUCHARIS.

Plate XXVII. fig. 4.

SPECIES. ANTHOCARIS EUCHARIS: alis rotundatis integerrimis albis; anticis apice fulvis margine nigro; posticis immaculatis seu punctis marginalibus nigris, his infra macula costali subferruginea. Expans. alar. unc. 1½.

ANTHOCARIS: with the wings rounded, entire, and white; the anterior fulvous at the tips, with a black margin; the posterior without spots, or with small marginal black points, the latter beneath, with a small reddish costal spot. Expanse of the wings, 1½ inch.

SYN. Papilio Eucharis, *Fabr. Syst. Ent. p.* 472. *Ent. Syst.* III. 1, *p.* 195. *Enc. Méth.* 9. 124. *Boisduval Hist. Nat. Lepid. p.* 568. (Anthocaris E.)
Papilio Aurora, *Cramer, pl.* 299, *f. A, B, C, D.*
Pieris Titea, *Enc. Méth.* 9. 124. *Horsfield Lep. Jav. p.* 141.

HABITAT. "India orientali," (*Fabricius*). Coromandel, Pegu, (*Boisduval*).

ANTHOCARIS GENUTIA.

Plate XXVII. fig. 5.

SPECIES. ANTHOCARIS GENUTIA: alis falcatis, integerrimis, albis; anticis apice fulvis; posticis punctis marginalibus nigris, subtus viridi marmoratis. Expans. alar. unc. 1½.

ANTHOCARIS: with the wings entire, white; the anterior falcate, with the tip fulvous; the posterior with marginal black spots, and on the under side marbled with green. Expanse of the wings, 1½ inch.

SYN. Papilio (Dan. Cand.) Genutia, *Fabricius Ent. Syst.* III. 1, *p.* 193. (nec Pap. Genutia Cramer, *pl.* 206, *f. C, D.*) *Enc. Méth.* 9, *p.* 168, (Pieris G.) *p.* 806, (Libythea G.) *Boisduval Hist. Nat. Lepid.* 1. 565.
Mancipium vorax Medea, *Hubn. Exot. Samml.*
♀ Pieris L'Herminieri, *Enc. Méth.* 9. 167.

There appears to have been two errors committed respecting this insect. In the first place, it is evidently very closely allied to the orange-tipped butterfly of our country, (Anthocaris Cardamines); the palpi, as represented in this figure, are fictitious, being elongated like those of the genus Libythea, which induced M. Godart to place it in that genus, in the appendix to his article on the butterflies in the Encyclopédie Méthodique. In the second place, instead of being an inhabitant of the East Indies, as stated by Fabricius, (and on his authority introduced into this work,) it is now known to belong to North America: a species agreeing in all

LEPIDOPTERA.

respects with the Fabrician insect, but with ordinary sized palpi, having been since discovered in North America, and described in the Encyclopédie Méthodique, under the name of Pieris L'Herminieri.

PIERIS AMARYLLIS.

Plate XXVIII. fig. 1.

SPECIES. PIERIS AMARYLLIS : alis rotundatis, integerrimis, concoloribus, obscure albis; anticis utrinque lunula media nigra. Expans. alar. unc. 2¾.
PIERIS : with the wings rounded, entire, concolorous, dirty white; the anterior pair having a small central black lunule visible both above and beneath. Expanse of the wings, 2¾ inches.
SYN. Papilio (Dan. Cand.) Amaryllis, *Fabr. Ent. Syst.* III. 1, *p.* 189. *Enc. Méth.* 9. 141. *Boisduval Hist. Nat. Lep.* 1. *p.* 549.

M. Boisduval expresses a doubt whether this species be really an inhabitant of the East Indies.

PIERIS CASTALIA.

Plate XXVIII. fig. 2.

SPECIES. PIERIS CASTALIA : alis integerrimis, rotundatis, albis, supra immaculatis subtus basi flavescentibus. Expans. alar. unc. 2¼.
PIERIS : with the wings entire, rounded, and white, without any spots above, but with the base of the wings yellowish beneath. Expanse of the wings, 2¼ inches.
SYN. Papilio (Dan. Cand.) Castalia, *Fabr. Ent. Syst.* III. 1. *p.* 188. *Enc. Méth.* 9. 160. (Pieris C.) *Boisduval Hist. Nat. Lepid. p.* 516.

COLIAS (CALLIDRYAS) SCYLLA.

Plate XXVIII. fig. 3.

GENUS. COLIAS, *Latreille et Godart.* (SUBGENUS : Callidryas, *Boisduval.*)
SPECIES. COLIAS (CALLIDRYAS) SCYLLA : alis integerrimis subrotundatis, supra anticis albis limbo nigro, posticis aurantiacis subtus omnibus nebulosis. Expans. alar. 2½.
COLIAS (CALLIDRYAS) : with the wings entire and rounded, above the anterior are white, with a black posterior border, and the posterior are rich orange; beneath all the wings are mottled with yellowish. Expanse of the wings, 2½ inches.

LEPIDOPTERA.

Syn. Papilio (Dan. Cand.) Scylla, *Linn. Syst. Nat.* 2. 763. *Fabr. Ent. Syst.* III, 1. p. 201. Cramer, tab. 12, f. C, D. (♂) Sulzer Gesch. Ins. t. 15, f. 6. *Enc. Méth.* 9. 95. Boisduval *Hist. Nat. Lep.* p. 631.
♂ Papilio Cornelia, *Fabr. Mant.* 2. p. 21.
Colias Scylla, *Horsfield Lep. Jav.* p. 133, pl. 4. f. 6, (larva and pupa.)

The caterpillar of this very distinct species, according to Dr. Horsfield, is very common in the eastern part of Java upon the Cassia fistula and obtusifolia. It is extremely abundant, particularly in the early part of the rainy season, after the renewal of the foliage of these plants. It is green, with minute black tubercles arranged in transverse series, and with a lateral yellow line above the legs. The chrysalis is boat-shaped, with the head pointed, but not so much elevated as some other species of this genus.

The last three species are represented on Dolichos lignosus.

VANESSA CACTA.

Plate XXIX. fig. 1.

Species. Vanessa Cacta: alis angulato-dentatis, anticis nigris basi purpureis macula magna transversa fulva, posticis fuscis strigis duabus submarginalibus. Expans. alar. unc. 2½.
Vanessa: with the wings angulato-dentated, the anterior black, with the basal portion purple, and with a large transverse orange spot; the posterior brown with two submarginal black lines. Expanse of the wings, 2½ inches.
Syn. Papilio (Nymphalis) Cacta, *Fabr. Ent. Syst.* III. 1, p. 116. *Enc. Méth.* 9. 309.

Originally described by Fabricius from the collection and drawings of Mr. Jones, from which Donovan most probably also obtained the present figure.

NYMPHALIS OCTAVIUS.

Plate XXIX. fig. 2.

Species. Nymphalis Octavius: alis nigris, fascia lata communi viridi, antice abbreviata, posticis caudatis; subtus griseis striga fusca. Expans. alar. unc. 2½.
Nymphalis: with black wings, a broad green bar running across them, not extending to the anterior margin of the anterior pair; posterior pair tailed; beneath grey with a brown streak. Expanse of the wings, 2½ inches.

LEPIDOPTERA.

Syn. Papilio (N.) Octavius, *Fabr. Ent. Syst.* III. 1. *p.* 73. (*nec* P. Octavia, *p.* 120.) *Enc. Méth.* 9. 368. (Nymphalis D.)

Fabricius gave India as the habitat of this species, but in the Encyclopédie Méthodique it is said to be from South America, being in that work doubtfully regarded as the female of the Brazilian species Nymphalis Iphis of Humboldt and Bonpland's voyage, which is therein considered as identical with the Papilio (N.) Morvus of Fabricius, by whom also India is given as the habitat of the last-named species.

NYMPHALIS (CHARAXES) ATHAMAS.
Plate XXIX. fig. 3.

Genus. Nymphalis, *Latreille.* Papilio, (Nymphalis p.) *Fabricius.* (Subgenus: Charaxes, *Boisduval.* Jasia, *Swainson.*)

Species. Nymphalis (Charaxes) Athamas: alis supra nigris, utrinque fascia media lata glauca subhyaliná, subtus lunulis ferrugineis marginata. Expans. alar. unc. 3.
Nymphalis: with the wings above black, on each side with a broad central glaucous subhyaline bar, which on the underside is margined with ferruginous lunules. Expanse of the wings, 3 inches.

Syn. Papilio (Equit. Ach.) Athamas, *Drury, vol.* 1, *pl.* 2, *fig.* 4. *Cramer Pap. pl.* 89. *f. C, D. Enc. Méth.* 9. 353. (Nymphalis A.)
Papilio Pyrrhus, *Donovan*, 1*st edit.* (excl. Syn. *Linn.*)

MORPHO MENETHO.
Plate XXX. fig. 1.

Genus. Morpho, *Fabricius, Latreille, Godart.*

Species. Morpho Menetho: alis dentatis fuscis, strigis duabus submarginalibus macularum flavarum, subtus flavescentibus, fasciis duabus fuscis. Expans. alar. unc. $3\frac{3}{4}$.
Morpho: with brown dentated wings, and with two rows of submarginal yellow spots, beneath yellowish with two brown fasciæ. Expanse of the wings, $3\frac{3}{4}$ inches.

Syn. Papilio (Nymph.) Menetho, *Fabr. Ent. Syst.* III. 1, *p.* 83. *Enc. Méth.* 9. 446.

This insect is nearly allied to Morpho Tullia, *Fabricius*. The present figure does not precisely agree with the Fabrician description; but as both were, in all probability, derived from the same source, Mr. Jones's collection of drawings, there can be but little doubt upon their identity.

LEPIDOPTERA.

HIPPARCHIA? ARCESILAUS.

Plate XXX. fig. 2.

GENUS. HIPPARCHIA, *Fabricius.* Satyrus, *Latreille, Godart.*

SPECIES. HIPPARCHIA? ARCESILAUS: alis integerrimis, supra ferrugineis immaculatis, subtus fuscis strigis duabus obscurioribus. Expans. alar. unc. 2$\frac{1}{8}$.
   HIPPARCHIA? with entire wings, above ferruginous immaculate, beneath brown with two darker parallel streaks. Expanse of the wings, 2$\frac{1}{8}$ inches.

SYN. Papilio (Nymph.) Arcesilaus, *Fabr. Ent. Syst.* III. 1. p. 153. *Enc. Méth.* 9. 497. (Satyrus? A.) (*nec* Pap. Arcesilaus, *Cramer, pl.* 294.)

HABITAT. "Siam, *Mus. Dom. Banks,*" (*Fabricius*).

Fabricius is silent regarding the submarginal row of white dots represented in this figure.

NYMPHALIS ORSIS.

♀ Plate XXX. fig. 3.

SPECIES. NYMPHALIS: alis nigris, (in mare cæruleo micantibus) strigis tribus macularibus, posticis striga marginali cærulea. Expans. alar. unc. 2$\frac{1}{2}$.
   NYMPHALIS: with the wings black, (in the male shining blue) with three transverse rows of white spots, and with a submarginal blue streak on the posterior wings. Expanse of the wings, 2$\frac{1}{2}$ inches.

SYN. Papilio (Nymph. Phal.) Orsis, *Drury Illustr. vol.* 3, *pl.* 16, *f.* 3.
   Papilio (N.) Orsis, *Fabr. Ent. Syst.* III. 1. p. 124. ♂. *Enc. Méth.* 9. 381.
   Papilio (N.) Blandina, *Fab. Ent. Syst.* III. 1. p. 129. ♀. *Donovan, 1st edit.*

This Brazilian insect was probably introduced into this work in consequence of the vague expression of Fabricius, "Habitat in Indiis," which was applied not only to N. Blandina, (or the female), but also to N. Orsis, (or the male.)

NYMPHALIS LIBERIA.

Plate XXX. fig. 4.

SPECIES. NYMPHALIS LIBERIA: alis dentatis, supra fulvis, anticarum arca apicis obscura; posticis supra puncto atro, subtus tribus ocellaribus. Expans. alar. unc. 2.
   NYMPHALIS: with the wings dentated, above fulvous, the anterior with a dark

LEPIDOPTERA.

coloured curved mark near the tips, the posterior with a single black dot above, and with three ocellated spots beneath. Expanse of the wings, 2 inches.

SYN. Papilio (Nymph.) Liberia, *Fabr. Ent. Syst.* III. 1. *p.* 135. *Enc. Méth.* 9. 375.

The same observation may be applied to this South American species as to the last,—Stoll, who investigated the transformations of many of the Lepidoptera of Guiana, having described its preparatory states. The caterpillar is of a dark green colour, with the head blue, and the legs and anal fork yellow. It is armed with black branching spines, of which the two anterior ones are by far the longest. The chrysalis is elongated, with the head bifid, and of a green colour. It appears to be a very variable species, as the Papilio Agatha and Merione of Fabricius, P. Laothoe of Herbst, and P. Ariadne of Cramer, are referred to it in the Encyclopédie Méthodique.

NYMPHALIS? PHEGEA.

Plate XXXI. fig. 1.

SPECIES. NYMPHALIS? PHEGEA: alis dentatis fuscis, nigro undatis, anticis fascia maculari, posticis disco ferrugineo aut albo; subtus pallidioribus. Expans. alar. unc. $3\frac{3}{4}$.

NYMPHALIS? with the wings dentated, brown, with numerous black waves; the anterior with a row of spots, and the disc of the posterior wings of a ferruginous or white colour; beneath the wings are paler, and the black waves more distinct. Expanse of the wings, $3\frac{3}{4}$ inches.

SYN. Papilio (N.) Phegea, *Fabr. Ent. Syst.* III. 1. *p.* 132. *Enc. Méth.* 9. 406, (Nymphalus? P.)

NYMPHALIS FATIMA.

Plate XXXI. fig. 2.

SPECIES. NYMPHALIS FATIMA: alis atris, fascia communi flava, posticarum abbreviata; anticis punctis quatuor flavis subapicalibus; posticis maculis rubris mediis, subcaudatis. Expans. alar. unc. $2\frac{1}{2}$.

NYMPHALIS: with the wings black, having a yellow bar running across them, which is abbreviated in the posterior pair; the anterior having also four subapical yellow spots, and the posterior several red spots in the middle, the latter slightly tailed. Expanse of the wings, $2\frac{1}{2}$ inches.

LEPIDOPTERA.

Syn. Papilio (Nymph.) Fatima, *Fabr. Ent. Syst.* III. 1. *p.* 81. *Enc. Méth.* 9. 375, (Nymphalis F.)

Fabricius gives the habitat of this species "in Indiis," with the observation, that it is of the figure of the South American Papilio Iphicla. (*Drury*, 1. *pl.* 14, *f.* 3, 4.) Hence I think it most probable that the term must be considered to signify the West Indies, or South America, and that the species is incorrectly introduced into the present work.

ARGYNNIS THYELIA.

Plate XXXI. fig. 3.

Genus. Argynnis, *Fabricius, Latreille, Godart.* Argyreus, *Scop.*

Species. Argynnis Thyelia : alis subrotundatis subdentatis, supra fulvis extimo maculisque nigris ; posticis subtus pallide testaceis fascia alba, transversa, media, punctis duobus baseos roseis nigroque cinctis. Expans. alar. $2\frac{1}{3}$ unc.

Argynnis : with the wings rather rounded and dentate, above fulvous, with the tip and several spots black ; the posterior beneath pale brick red, with a white bar across the middle, and with two crimson basal spots, bordered with black. Expanse of the wings, $2\frac{1}{3}$ inches.

Syn. Papilio (N.) Thyelia, *Fabr. Ent. Syst.* III. 1. *p.* 142. *Enc. Méth.* 9. 257. (Argynnis T.)

PIERIS NERO.

Plate XXXII. fig. 1.

Species. Pieris Nero : alis integerrimis, anticis elongato-trigonis, sanguineis, margine parum fuscescente, subtus aurantiis ; posticis strigis duobus fuscis fere obsoletis. Expans. alar. unc. 3.

Pieris : with the wings entire, the anterior elongate-trigonate, sanguineous, with the nerves and margin brownish, beneath orange ; the posterior wings with two nearly obsolete streaks. Expanse of the wings, 3 inches.

Syn. Papilio (Nymph.) Nero, *Fabricius Ent. Syst.* III. 1. *p.* 153. *Enc. Méth.* 9. 805, *Supp.* (Pieris N.) *Boisduval Hist. Nat. Lepid.* 1. 485.

Pieris Thyria, *Enc. Méth.* 9, 147. *Guerin Icon. R. An. Ins.* pl. 77, *f.* 1. *Horsfield, Zool. Jour.* vol. 4, *pl.* 4, *f.* 2.

Habitat. "In Asia, *Mus. Britann.*" (*Fabricius*). Java, (*Horsfield, Boisduval*).

LEPIDOPTERA.

NYMPHALIS GNIDIA.

Plate XXXII. fig. 2.

SPECIES. NYMPHALIS GNIDIA: alis dentatis testaceis; anticis apice fuscis fascia media punctisque subapicalibus albis; posticis striga fulva, lunulis nigris. Expans. alar. unc. 2½.

NYMPHALIS: with the wings dentated testaceous; the anterior with the extremity brown, with a white central fascia, and several subapical white spots; the posterior with a submarginal fulvous band with black lunules. Expanse of the wings, 2½ inches.

SYN. Papilio (N.) Gnidia, *Fabr. Ent. Syst.* III. 1, *p.* 137. *Enc. Méth.* 9. 386 ? (Nymphalis G. ?)

HABITAT. "In Indiis," (*Fabricius*).

Described by Fabricius from the collection and drawings of Mr. Jones, which also, in all probability, supplied the accompanying figure.

BIBLIS HIARBAS.

Plate XXXII. fig. 3.

GENUS. BIBLIS, *Fabricius, Latreille, Godart.* Papilio, *Donovan.*

SPECIES. BIBLIS HIARBAS: alis dentatis, fuscis; fascia utrinque communi alba, posticarum latiore, anticarum abbreviata; marginibus lunulis rufis. Expans. alar. unc. 2½.

BIBLIS: with the wings dentated, brown, with a white fascia running through the wings both above and beneath; being narrower and abbreviated in the anterior pair; the margins of the wings with red lunules. Expanse of the wings, 2½ inches.

SYN. Papilio (N.) Hiarbas, *Drury Illustr.* vol. 3, *pl.* 14. *Fabr. Ent. Syst.* III. 1, *p.* 128. (Papilio (N.) Hiarba). *Enc. Méth.* 9. 824. (Biblis Hiarba.)

Drury, to whom Fabricius was indebted for his knowledge of this species, states that he received it from Sierra Leone. Fabricius, however, gave his usual vague locality, "In Indiis," upon which authority Donovan introduced it into this work. It is, however, a widely distributed African insect, being found in the country of the Hottentots, (*Enc. Méth.*) and having been received by me from the Cape of Good Hope.

LEPIDOPTERA.

NYMPHALIS ISIS.

Plate XXXIII. fig. 1.

SPECIES. NYMPHALIS ISIS: alis fusco-nigris, anticis integris, utrinque macula disci chermesina; posticis dentatis supra striga marginali albida; omnibus subtus viridilineatis. Expans. alar. unc. 2½.

NYMPHALIS: with blackish brown wings; the anterior entire with a large discoidal sanguineous spot; the posterior dentated, with a whitish marginal streak; all the wings beneath with green lines. Expanse of the wings, 2½ inches.

SYN. Papilio (Dan. Fest.) Isis, *Drury*, vol. 3, pl. 7, f. 1, 2. *Fab. Ent. Syst.* III. 1. p. 124. *Enc. Méth.* 9. 421. (Nymphalis I.)

The Fabrician expression, "In Indiis," indicating the habitat of this species, has evidently again induced our author to give this insect a place in these Illustrations. That it is an inhabitant of Brazil, we have the authority both of Drury and the Encyclopédie Méthodique.

NYMPHALIS PHORCYS.

Plate XXXIII. fig. 2.

SPECIES. NYMPHALIS PHORCYS: alis dentato-subcaudatis, supra fuscis immaculatis, subtus obscurius strigosis; posticis punctis duobus cinereis. Expans. alar. unc. 2¾.

NYMPHALIS: with the wings dentate, slightly tailed, the upper side brown immaculate, beneath with several obscure slender streaks; the posterior also with two ashy-coloured spots. Expanse of the wings, 2¾ inches.

SYN. Papilio (N.) Phorcys, *Fabr. Ent. Syst.* III. 1. p. 80. *Enc. Méth.* 9. 372.

NYMPHALIS ERIBOTES.

Plate XXXIII. fig. 3.

SPECIES. NYMPHALIS ERIBOTES: alis subcaudatis, fulvis aut ferrugineis, basi violaceo micantibus, subtus griseis. Expans. alar. unc. 2¾.

NYMPHALIS: with the wings slightly tailed, above fulvous or ferruginous, with the base shining with violet or purple; all the wings beneath greyish. Expanse of the wings, 2¾ inches.

SYN. Papilio (N.) Eribotes, *Fabr. Syst. Ent.* p. 484. *Ent. Syst.* III. 1. p. 73. *Enc. Méth.* 9. 316, (Nymphalis E.)

LEPIDOPTERA.

Papilio Leonida, *Cramer, pl.* 338, *f. C, D. Stoll Suppl. Cramer, pl.* 6, *f.* 2, *A, B.* (larva et pupa.)

Fabricius described this insect from the collection of Dr. Hunter, now belonging to the University of Glasgow, and gave its habitat " In India." It is, however, a native of Guiana, in South America, Stoll having reared it from the caterpillar state in that country. The caterpillar is black, with short hairs, and a great number of white dots. The chrysalis is short, without angular projections, of a grey colour, and is found suspended by its tail from the twigs of the shrubs upon which the larva had been nourished. In the description given in the Encyclopédie Méthodique, the anal angle of the posterior wings is described as being ornamented with four or five small black and white spots, of which no mention is made by Fabricius, nor are they indicated in the accompanying figure.

NYMPHALIS ISIDORE.

Plate XXXIII. fig. 4.

SPECIES. NYMPHALIS ISIDORE: alis anticis falcatis, posticis caudatis; omnibus fulvis, anticis punctis duobus mediis pallidis, macula costali apiceque fuscis. Expans. alar. unc. 3.

NYMPHALIS: with the anterior wings falcate, the posterior tailed; all the wings above fulvous, the anterior with two pale spots on the middle, and with a costal spot, and the extremity brown. Expanse of the wings, 3 inches.

SYN. Papilio Isidore, *Cramer Pap. pl.* 235, *fig. A, B, E, F. Herbst. Pap. t.* 150, *f.* 1, 2. *Enc. Méth.* 9. 371. (Nymphalis I.)

Fabricius again uses the expression " In Indiis," to designate the habitat of this species, which, in the Encyclopédie Méthodique is stated to be an inhabitant of Guiana and Brazil.

CETHOSIA CYDIPPE.

♂ Plate XXXIV. fig. 1.

GENUS. CETHOSIA, *Fabricius, Latreille, Godart.*

SPECIES. CETHOSIA CYDIPPE: alis dentatis, basi rufis, apice nigris, lunulis subapicalibus serie duplici digestis albis; anticis fascia lata abbreviata pone medium; subtus basi testaceis nigro cæruleoque variis. Expans. alar. unc. 4¼.

CETHOSIA: with the wings dentated, the base red, and the tips black, with white lunules arranged in a double series within the exterior margins; the anterior

LEPIDOPTERA.

having also a broad short white fascia beyond the middle; beneath at the base testaceous, with black and blue markings. Expanse of the wings, $4\frac{1}{4}$ inches.

SYN. Papilio Cydippe, *Linn. Syst. Nat.* 2. 776. *Fabr. Ent. Syst.* III. 1. *p.* 112. *Clerck Icon. tab.* 36, *f.* 1. *Herbst. Pap. t.* 245, *f.* 4, 5.
Papilio Iris, *Cramer Pap. pl.* 62, *fig. A, B.*

HABITAT. "In India," (*Linnæus.*) In China, (*Enc. Méth.*)

NYMPHALIS DIRCE.

Plate XXXIV. fig. 2.

SPECIES. NYMPHALIS DIRCE: alis fuscis, anticis utrinque fascia flava; subtus omnibus albido nigroque striatis; posticis angulatis. Expans. alar. unc. 2—3.
NYMPHALIS: with the wings brown, the anterior on both sides with a broad oblique yellow band; beneath all the wings striped with pale buff and brownish black; the posterior wings angulated. Expanse of the wings, 2 to 3 inches.

SYN. Papilio Dirce, *Linn. Syst. Nat.* 2. 778. *Cramer, pl.* 212, *f. C, D. Stoll Suppl. Cramer, pl.* 2, *f.* 3, *A, B,* 4, *A, B. Fabr. Ent. Syst.* III. 1. *p.* 123. *Enc. Méth.* 9. 371, (Nymphalis D.)

Linnæus and Fabricius state of this species, "Habitat in Indiis." It is, however, an inhabitant of Guiana and Brazil, Stoll having reared it in that country; the caterpillar is black.

NYMPHALIS EURINOME.

Plate XXXIV. fig. 3.

SPECIES. NYMPHALIS EURINOME: alis subdentatis, nigris, maculis permultis posticarumque disco baseos albis. Expans. alar. unc. $3\frac{3}{4}$.
NYMPHALIS: with the wings slightly dentate, black, with a considerable number of spots, and the basal disc of the posterior wings white. Expanse of the wings, $3\frac{3}{4}$ inches.

SYN. Papilio Eurinome, *Cramer Pap. tab.* 70, *fig. A. Fabr. Ent. Syst.* III. 1. *p.* 57. *Enc. Méth.* 9. 398.

HABITAT. In India orientali.

NYMPHALIS HIPPONA.

Plate XXXV. fig. 1.

Species. Nymphalis Hippona: alis supra nigris; anticis fulvo flavoque variis; posticis caudatis, basi fulvo, apice nigro striga marginali lunularum albarum. Expans. alar. unc. 4.

Nymphalis: with the wings black above; the anterior varied with fulvous and yellow; the posterior with a long tail, the base fulvous, the apex black, with a row of marginal white lunules. Expanse of the wings, 4 inches.

Syn. Papilio (N.) Hippona, *Fabr. Sp. Ins.* 2, p. 54. *Enc. Méth.* 9. 362. *Drury Illustr.* (2nd edit.) vol. 3, p. 21.
Papilio (N.) Fabius, *Cramer Pap.* pl. 90, fig. C, D. *Stoll Suppl. Cramer*, pl. 2, f. 1, 1 A, 1 B, 1 C. *Drury Illustr.* (1st edit.) vol. 3, pl. 16, f. 1, 2.

Fabricius erroneously gave "India" as the habitat of this curious insect, which induced Donovan to introduce it into this work. It is, however, a native of Guiana and Brazil. Its transformations have been observed by Stoll, who has figured the insect in its different states, as above referred to. The caterpillar is dark green, with a black dorsal line and lateral black spots; the head is armed with two short obtuse spines. It feeds only by night, concealing itself by day in a rolled-up leaf; the chrysalis is short and thick, without angular projections.

CETHOSIA CYANE.

Plate XXXV. fig. 2.

Species. Cethosia Cyane: alis dentatis, nigris, linea tenuissima angulata submarginali, anticis fascia, posticis disco (nigro-punctato) albis. Expans. alar. unc. 3½.

Cethosia: with the wings dentate, black, with a very slender angulated line running within the posterior margin of the wings, the anterior having a fascia, and the posterior the disc white, the latter spotted black. Expanse of the wings, 3½ inches.

Syn. Papilio (N. P.) Cyane, *Drury, App.* vol. 2, 1st edit. *Herbst. Pap.* tab. 248, f. 3, 4. *Cramer Pap.* pl. 295, f. C, D. *Fabr. Ent. Syst.* III. 1. p. 115.

Habitat. Bengal, (*Drury*). India, (*Fabricius*).

LEPIDOPTERA.

NYMPHALIS CÆNOBITA.

Plate XXXV. fig. 3.

SPECIES. NYMPHALIS CÆNOBITA : alis dentatis, nigris, anticis stria maculisque, posticis supra fascia alba; subtus albis, fasciis quatuor maculisque marginalibus fuscis. Expans. alar. unc. 2½.

NYMPHALIS : with the wings dentate, black ; the anterior with a broad band, and numerous spots and lunules of white ; the posterior above with a white fascia ; beneath white with four fasciæ, and marginal spots of brown. Expanse of the wings, 2½ inches.

SYN. Papilio (N.) Cænobita, *Fabr. Ent. Syst.* III. 1. *p.* 247. *Enc. Méth.* 9. 433. (Nymphalis C.)

HABITAT. "In Indiis," (*Fabricius*).

NYMPHALIS (ACONTHEA) COCALIA.

Plate XXXVI. fig. 1.

GENUS. NYMPHALIS, *Latreille, Godart.* (SUBGENUS : Aconthea, *Horsfield.*)

SPECIES. NYMPHALIS (ACONTHEA) COCALIA : alis dentatis, fuscis ; anticis nigro flavoque maculatis ; subtus omnibus griseis striga punctorum alborum. Expans. alar. unc. 3.

NYMPHALIS (ACONTHEA) : with the wings dentated, brown ; the anterior varied with black and buff spots ; beneath all the wings are grey, with a row of white spots. Expanse of the wings, 3 inches.

SYN. Papilio (N.) Cocalia, *Fabr. Ent. Syst.* III. 1. *p.* 250. *Enc. Méth.* 9. 405. (Nymphalis C.) *nec* Papilio Cocala, *Herbst. et Cramer.*

HIPPARCHIA BALDUS.

Plate XXXVI. fig. 2.

SPECIES. HIPPARCHIA BALDUS : alis integerrimis, fuscis, subtus cinereo undatis ; anticis utrinque ocello magno pupilla gemina ; posticis supra ocellis quatuor, subtus sex. Expans. alar. unc. 1½.

HIPPARCHIA : with the wings entire, brown, beneath with grey waves ; the anterior on each side with a large ocellus, having a double pupil ; the posterior above, with four, beneath with six eyelets. Expanse of the wings, 1½ inch.

SYN. Papilio (Satyrus) Baldus, *Fabr. Syst. Ent* III. 1. *p.* 223. *Enc. Méth.* 9. 551. (Satyrus B.)

Papilio Lysandra, *Cramer, pl.* 293, *fig.* G, H.

LEPIDOPTERA.

VANESSA SOPHIA.

Plate XXXVI. fig. 3.

SPECIES. VANESSA SOPHIA: alis denticulatis, anticis subfalcatis, posticis rotundatis, omnibus supra fulvo, nigro, flavidoque variis, limbo fusco nigro punctata, lunulis posticarum albis. Expans. alar. unc. 1¾—2.

VANESSA: with the wings denticulated, the anterior subfalcate, the posterior rounded above; all the wings are varied with fulvous, black, and buff, with a broad brown border with black spots, and with submarginal white lunules in the posterior wings. Expanse of the wings, 1¾ to 2 inches.

SYN. Papilio (Satyrus) Sophia, *Fabr. Ent. Syst.* III. 1. *p.* 248. *Enc. Méth.* 9. 823.

Fabricius described this species from the collection of Mr. Drury, and the drawings of Mr. Jones, with the habitat "In Indiis." In the Encyclopédie Méthodique, the western coast of Africa is stated to be its native country.

NYMPHALIS AUGE.

Plate XXXVI. fig. 4.

SPECIES. NYMPHALIS AUGE: alis dentatis, fuscis; anticis fasciis tribus viridibus; posticis fulvis serie submarginali punctorum nigrorum. Expans. alar. unc. 1⅘.

NYMPHALIS: with the wings dentated, brown; the anterior with three greenish fasciæ, the posterior fulvous with a submarginal row of black spots. Expanse of the wings, 1⅘ inch.

SYN. Papilio (Satyrus) Auge, *Fabr. Ent. Syst.* III. 1. *p.* 248. *Enc. Méth.* 9. 387.

This insect is also stated by Fabricius to have been in the collection of Mr. Drury, and figured by Mr. Jones, and the habitat is also given "In Indiis." From its near relationship, however, with N. Sophia and N. Doriclea, both of which are inhabitants of western tropical Africa, I have little doubt that N. Auge is from that quarter of the globe, and not from either of the Indies.

NYMPHALIS PERIANDER.

Plate XXXVII. fig. 1.

SPECIES. NYMPHALIS PERIANDER: alis utrinque albis, strigis sex flavescentibus, margine postico fusco albo strigoso; posticis caudatis. Expans. alar. unc. 1¾.

LEPIDOPTERA.

NYMPHALIS: with the wings on both sides white with six yellow stripes, posterior margin brown with white lines; posterior wings tailed. Expanse of the wings, 1¾ inch.

SYN. Papilio (N.) Periander, *Fabr. Ent. Syst.* III. 1. *p.* 67. *Enc. Méth.* 9. 362. (Nymphalis P.) *Horsfield Lep. Jav. pl.* 5, *f.* 3, *and* 3 *a.* (Paphia P.)

HABITAT. Mysore, (*Donovan*). Java, (*Enc. Méth.*)

ERYCINA (ZEMEROS) ALLICA.
Pl. XXXVII. fig. 2.

FAMILY. LYCÆNIDÆ, *Leach*.

GENUS. ERYCINA, *Fabricius, Latreille, Godart*. (SUBGENUS: Zemeros, *Boisduval*.)

SPECIES. ERYCINA (ZEMEROS) ALLICA: alis denticulatis obscure fulvis, punctis, nigris numerosis, albo, fœtis. Expans. alar. unc. 1½.

ERYCINA (ZEMEROS): with denticulated wings of an obscure fulvous colour, with numerous black spots, accompanied with smaller white ones. Expanse of the wings, 1½ inch.

SYN. Papilio (S.) Allica, *Fabr. Ent. Syst.* III. 1. *p.* 244. *Enc. Méth.* 9. 567. (Erycina A.) *Boisduval Hist. Nat. Lepid.* 1. *pl.* 21, *fig.* 5. (Zemeros A.)

HABITAT. Siam, (*Fabricius*). China, Bengal, Java, (*Enc. Méth.*)

NYMPHALIS ANCÆAS.
♂ Pl. XXXVII. fig. 3.

SPECIES. NYMPHALIS ANCÆAS: alis supra atris (♂), aut fuscis, (♀); anticarum fascia cyanea obliqua, posticarum (♂) macula magna discoidali rufa; subtus viridibus, posticarum strigis tribus ferrugineis. Expans. alar. unc. 2½.

NYMPHALIS: with the wings above black in the male, or brown in the female; the anterior with a broad oblique cyaneous or green bar, the posterior in the male with a large discoidal red spot; beneath green, the posterior with three ferrugineous streaks. Expanse of the wings, 2½ inches.

SYN. ♂ Papilio (N.) Ancæa, *Linn. Syst. Nat.* 2. 781. *Fabr. Ent. Syst.* III. *p.* 154. *Enc. Méth.* 9. 409. *Cramer Pap. pl.* 338, *fig.* C, D.
Papilio Obrinus, *Donovan*, 1*st edit.*

♀ Papilio (N.) Obrinus, *Linn. Syst. Nat.* 2. *p.* 776. *Fabr. Ent. Syst.* III. 1. *p.* 154. *Cramer Pap. pl.* 49, *fig.* E, F. *Stoll Suppl. Cramer, pl.* 6. *f.* 5.

Stoll, by rearing this insect in Guiana, clearly proved not only the incorrectness of the old habitat assigned to it of India, but also that the two Linnæan species,

LEPIDOPTERA.

P. Ancæa and Obrinus, were the sexes of the same species. The caterpillar, according to the first named author, is green, with a lateral line, the head ferruginous, and armed with two branching spines; its back is also armed with shorter spines. The chrysalis is grey, and is suspended by the tail.

HIPPARCHIA CRANTOR.
Pl. XXXVII. fig. 4.

SPECIES. HIPPARCHIA CRANTOR : alis integerrimis, fuscis ; supra anticis immaculatis ; posticis ocello unico bipupillato ; subtus quinque, primo quartoque bipupillato. Expans. alar. unc. 1⅗.

HIPPARCHIA : with the wings entire, brown ; the anterior immaculate above, the posterior with a single ocellus with two pupils ; beneath with five ocelli, the first and fourth having two pupils. Expanse of the wings, 1⅔ inch.

SYN. Papilio, (N.) Crantor, *Fabr. Ent. Syst.* III. 1. *p.* 158. *Enc. Méth.* 9. 488. (Satyrus C.)

HABITAT. "India," *Fabricius.* Brazil, (*Enc. Méth.*)

NYMPHALIS LIRISSA.
Pl. XXXVII. fig. 5.

SPECIES. NYMPHALIS LIRISSA : alis subdentatis, cinereis, fusco undatis ; anticis fascia alba ; posticis punctis quatuor ocellaribus albis. Expans. alar. unc. 1¾.

NYMPHALIS : with the wings slightly dentated, cinereous, with brown waves ; the anterior with a white bar, and the posterior with four white eyelets. Expanse of the wings, 1¾ inch.

SYN. Nymphalis Lirissa, *Enc. Méth.* 9. 406.
Papilio (S.) Liria, *Fabr. Ent. Syst.* III. 1, *p.* 239, (*nec* Papilio (N.) Liria, *Fabr. Ent. Syst.* III. 1. *p.* 126.)

HABITAT. "In Indiis," (*Fabricius.*) Brazil, (*Enc. Méth.*)

THECLA ISOCRATES.
♀ Pl. XXXVIII. fig. 1.

FAMILY. LYCÆNIDÆ, *Leach, Stephens.*

GENUS. THECLA, *Fabricius.* (*Syst. Gloss.*) Polyommatus p. *Latreille, Godart.* Cupido p. *Schrank.*

LEPIDOPTERA.

SPECIES. THECLA ISOCRATES: alis supra fuscis, (♂ cœruleo-micantibus, ♀ macula fulva); posticis (♀) macula submarginali; subtus cinereis, litura gemina strigaque duplici nigricantibus, anguloque ani ocellis duobus. Expans. alar. unc. 1¾.

THECLA : with the wings above brown, in the male shining blue, in the female with a fulvous patch ; the posterior in the female with a submarginal eyelet ; beneath cinereous, with a double mark and double streak of blackish, the anal angle with two eyelets. Expanse of the wings, 1¾ inch.

SYN. Papilio (H. R.) Isocrates, *Fabr. Ent. Syst.* III. 1. *p.* 266. *Enc. Méth.* 9. 633.
Papilio (H. R.) Pann, *Fabr. Ent. Syst.* III. 1. *p.* 276. *Donovan, 1st edit.* (*nec* Thecla Pan,) *Drury*, 2. *pl.* 23, *fig.* 3, 4.

HABITAT. India, (*Fabricius*). Bengal, (*Enc. Méth.* 9.)

THECLA PINDARUS.

Plate XXXVIII. fig. 2.

SPECIES. THECLA PINDARUS : alis subtricaudatis, supra cœruleis, limbo atro ; subtus fuscis argenteo fulvoque maculatis. Expans. alar. unc. 1½.

THECLA : with three short tails ; wings above blue with a black border, beneath varied with silvery and fulvous bars. Expanse of the wings, 1½ inch.

SYN. Hesperia (R.) Pindarus, *Fabr. Ent. Syst.* III. 1. *p.* 262.
Polyommatus Vulcanus (var. ♂) *Enc. Méth.* 9. 644.

Donovan observes that " Fabricius describes this beautiful insect as a native of India, and for its figure refers only to the original drawings of W. Jones, Esq. We have ascertained the species from those drawings, collated with the manuscript in the hand-writing of Fabricius, and on this authority give it a place in our selection of Indian insects."

I have hesitated in adopting the opinion of the editors of the Encyclopédie Méthodique, that this is an extraordinary variety of Thecla Vulcanus.

THECLA VULCANUS.

Plate XXXVIII. fig. 3.

SUBGENUS. AMBLYPODIA, *Horsfield.*

SPECIES. THECLA (A.) VULCANUS : alis bicaudatis, supra fuscis ; anticis fasciis quatuor, posticis margine fulvis, subtus strigis fulvis argenteisque variegatis. Expans. alar. unc. 1¼.

THECLA (A.) : with the wings with two tails, above brown ; the anterior with four short unequal transverse waved bars, and the posterior with the margin fulvous, beneath varied with fulvous and silvery bars. Expanse of the wings, 1¼ inch.

LEPIDOPTERA.

SYN. Hesperia (R.) Vulcanus, *Fabr. Ent. Syst.* III. 1. *p.* 264. *Enc. Méth.* 9. 644. Polyommatus V. ♀. *Horsfield Lep. Jav. p.* 106, (Amblypodia V.)
Papilio Etolus, *Cramer, pl.* 208, *f. E, F.*

Donovan states that the accompanying figures represent the male insect, the female being rather larger, with the colours on the underside more obscure, and the silvery stripes broader. The authors of the Encyclopédie Méthodique describe the male as possessing a shining violet tinge on the upper side, whence the accompanying figures must be females, and the supposed female rather a dark variety of their sex.

THECLA CHITON.

Plate XXXIX. fig. 1.

SPECIES. THECLA CHITON: alis cæruleis, limbo fusco; posticis tricaudatis, (interno parvo) striga marginali alba; subtus flavescenti-albis, nigroque fasciatis. Expans. alar. 1⅜ unc.

THECLA: with the wings blue, with a black border; the posterior with three tails, the interior very short, and with a submarginal white line; beneath yellowish white, with black bars. Expanse of the wings, 1⅜ inch.

SYN. Hesperia (R.) Chiton, *Fabr. Ent. Syst.* III. 1. *p.* 262.
Papilio Silenus, *Cramer Ins. t.* 282, *f. C, D.*
Papilio Agis, *Drury, 1st edit. vol.* 3, *pl.* 26, *f.* 3, 4.
Papilio (Pl. Urb.) Phaleros? *Linn. Syst. Nat.* 2. 796. *Enc. Méth.* 9. 629, (Polyommatus Ph.)

HABITAT. India, (*Linnæus, Fabricius*). Brazil, (*Drury*). Surinam, (*Enc. Méth.*)

In the Encyclopédie Méthodique this figure, and that of Drury, (*vol.* 3, *pl.* 26.) are considered as the Papilio Phaleros of Linnæus. The insect here represented is a male, according to the distinctions given in the Encyclopédie, but it has not the large orbicular brown spot near the middle of the anterior wings, so that it is, perhaps, doubtful whether the P. Phaleros, *Linnæus*, and the P. Chiton, *Fabricius*, be really identical, as Donovan had suggested.

THECLA HERODOTUS.

Plate XXXIX. fig. 2.

SPECIES. THECLA HERODOTUS: alis caudatis, cæruleis, subtus viridibus; posticis striga submarginali punctorum, extus alborum et intus nigrorum. Expans. alar. 1⅛ unc.

LEPIDOPTERA.

THECLA : with the wings tailed, above blue, beneath green; the posterior pair with a submarginal row of dots, white without and black within. Expanse of the wings, $1\frac{1}{6}$ inch.

SYN. Hesperia (R.) Herodotus, *Fabr. Ent. Syst.* III. 1. p. 286. *Enc. Méth.* 9. 641, (Polyommatus H.)
Papilio Amyntor, *Cramer, pl.* 48, *fig. E.*
♀ Papilio Eryx, *Linn. Mantissa,* 1. 537. *Fabr. Ent. Syst.* III. 1. p. 283.
Papilio Menalcas, *Cramer, pl.* 259, *fig. A, B.*

HABITAT. "In Indiis," (*Fabricius*). Surinam, (*Enc. Méth.*)

THECLA PYTHAGORAS.

Plate XXXIX. fig. 3.

SPECIES. THECLA PYTHAGORAS: alis tricaudatis, atris, disco fulvo, lunulisque submarginalibus; posticarum fulvis; subtus nigris albo variis fasciaque media alba, punctisque duobus irroratis ad angulum ani. Expans. alar. unc. $1\frac{1}{6}$.

THECLA: with the wings three-tailed, black, with the disc of each fulvous; the posterior with submarginal fulvous spots; beneath black varied with white; with a white central bar, and with two eyes at the anal angle. Expanse of the wings, $1\frac{1}{6}$ inch.

SYN. Hesperia (R.) Pythagoras, *Fabr. Ent. Syst.* III. 1. p. 259. *Enc. Méth.* 9. 619, (Polyommatus P.)
Var. (caudis mutilis) Hesperia (R.) Juba, *Fabr. Ent. Syst.* III. 1. p. 314.

Fabricius states the habitat of Pythagoras to be "In Indiis," but that of Juba, which is a variety of the same species, with the tails broken off, to be Sierra Leone. In the Encyclopédie Méthodique, the western coast of Africa is given as the habitat of Pythagoras.

POLYOMMATUS FLORUS.

Plate XXXIX. fig. 4.

GENUS. POLYOMMATUS, *Latreille.*

SPECIES. POLYOMMATUS FLORUS: alis integerrimis, fulvis, margine nigro; anticis supra punctis duobus; subtus omnibus basi punctis plurimis nigris. Expans. alar. unc. 1.

POLYOMMATUS: with the wings entire, fulvous, with a slender black margin; the anterior above with two black spots; beneath all the wings at the base spotted with black. Expanse of the wings, 1 inch.

SYN. Hesperia (R.) Florus, *Fabr. Ent. Syst.* III. 1. p. 310.

HABITAT. "In Indiis," (*Fabricius*).

LEPIDOPTERA.

THECLA LISIAS.

Plate XL. fig. 1.

SPECIES. THECLA LISIAS : alis tricaudatis ; anticis fuscis macula magna fulva ; posticis supra fuscis, margineque postico cæruleo maculato ; subtus albis nigro-maculatis. Expans. alar. unc. $1\frac{1}{4}$.
    THECLA : with three tails ; the anterior wings brown, with a large fulvous spot ; the posterior above brown, with the posterior margin spotted with blue ; beneath white with black marks. Expanse of the wings, $1\frac{1}{4}$ inch.

SYN. Hesperia (R.) Lisias, *Fabr. Ent. Syst.* III. 1. p. 261.

HABITAT. " Poulicandor, *Mus. Dom. Banks*," (*Fabricius*).

THECLA SOPHOCLES.

Plate XL. fig. 2.

SPECIES. THECLA SOPHOCLES : alis bicaudatis, nigris, disco communi cæruleo ; subtus albis, strigis undatis flavescentibus ; posticis puncto anali fulvo. Expans. alar. unc. $1\frac{1}{2}$.
    THECLA : with two tails, black, with the disc blue ; beneath white with several waved yellowish lines ; the posterior pair with an anal fulvous eyelet. Expanse of the wings, $1\frac{1}{2}$ inch.

SYN. Hesperia (R.) Sophocles, *Fabr. Ent. Syst.* III. 1, p. 267. *Enc. Méth.* 9. 631, (Polyommatus S.)

HABITAT. " In Indiis," (*Fabricius*).

THECLA JARBAS.

Plate XL. fig. 3.

SPECIES. THECLA JARBAS : alis caudatis, fulvis, limbo fusco ; subtus cinereis striga alba ; posticis punctis duobus atris. Expans. alar. unc. $1\frac{1}{2}$.
    THECLA : with the wings tailed, fulvous, with the border brown ; beneath cinereous with a white bar ; the posterior pair with two black spots. Expanse of the wings, $1\frac{1}{2}$ inch.

SYN. Hesperia (R.) Jarbas, *Fabr. Ent. Syst.* III. 1, p. 276. *Enc. Méth.* 9. 646. *Horsfield Lep. Jav. p.* 93, (Thecla J.)

HABITAT. "Siam, *Mus. Dom. Banks*," (*Fabricius*). Bengal and Java, (*Enc. Méth.*)

LEPIDOPTERA.

THECLA THALES.

Plate XL. fig. 4.

SPECIES. THECLA THALES: alis bicaudatis, utrinque atris, subtus lunulis cæruleis; posticis fascia submarginali aurea. Expans. alar. unc. 1¼.

THECLA: with two tails, the wings on both sides black, beneath with blue lunules; the posterior also with an abbreviated submarginal golden fascia. Expanse of the wings, 1¼ inch.

SYN. Hesperia (R.) Thales, *Fabricius Ent. Syst.* III. 1, *p.* 268. *Enc. Méth.* 9. 625, (Polyommatus T.)

HABITAT. "In Indiis," (*Fabricius*). Brazil, (*Enc. Méth.*)

THECLA MELIBŒUS.

Plate XLI. fig. 1.

SPECIES. THECLA MELIBŒUS: alis bicaudatis, cærulescentibus, limbo fusco; subtus flavescentibus; anticis fusco, posticis nigro strigosis, angulo ani atro; annulis cæruleis. Expans. alar. unc. 1⅓.

THELCA: with two tails, the wings blue above with a brown border; beneath yellowish, the anterior with brown and the posterior with black streaks; anal angle black with two blue rings. Expanse of the wings, 1⅓ inch.

SYN. Hesperia (R.) Melibœus, *Fabr. Ent. Syst.* III. 1. *p.* 271. *Enc. Méth.* 9. 629, (Polyommatus M.)

HABITAT. "In India, (*Fabricius*). Brazil, (*Enc. Méth.*)

THECLA TYRTÆUS.

Plate XLI. fig. 2.

SPECIES. THECLA TYRTÆUS: alis bicaudatis, fuscis; posticis subtus striga undata albā lunulisque submarginalibus nigris, intermediis rufis. Expans. alar. unc. 1.

THECLA: with two tails, wings above brown; beneath the posterior have a waved white streak and black submarginal lunules, the two middle lunules being red. Expanse of the wings, 1 inch.

SYN. Hesperia (R.) Tyrtæus, *Fabr. Ent. Syst.* III. 1. *p.* 271. *Enc. Méth.* 9. 637, (Polyommatus T.)

HABITAT. "In India," (*Fabricius*).

LEPIDOPTERA.

THECLA XENOPHON.

Plate XLI. fig. 3.

SPECIES. THECLA XENOPHON: alis bicaudatis fuscis, disco flavo; subtus cinereo, striga media alba fusco strigaque innata; posticarum marginali fusca. Expans. alar. unc. $1\frac{1}{8}$.

THECLA: with two tails, wings brown, with a yellow disc, beneath cinereous, with a central white and brown streak; the posterior also with a brown submarginal line. Expanse of the wings, $1\frac{1}{8}$ inch.

SYN. Hesperia (R.) Xenophon, *Fabr. Ent. Syst.* III. 1. p. 272. *Enc. Méth.* 9. 640. (Polyommatus X.) *Horsfield Lep. Jav.* p. 93, pl. 4, f. 2, 2a, (larva et pupa).
Papilio Melampus, *Cramer*, pl. 362, f. G, H.

HABITAT. "In India," (*Fabricius*). Java, (*Horsfield*).

Dr. Horsfield has traced the metamorphoses of this species, and pointed out the distinction between it and Jarbas, plate XL. fig. 3. The caterpillar is more elongate than the ordinary onisciform larvæ, with short stumps of fascicles. It varies in colour from yellow, with a greenish cast, to a dark ferruginous brown, with lateral bands. It feeds upon Schmiedelia racemosa, and was found in considerable abundance by Dr. Horsfield in Java.

THECLA ACHÆUS.

Plate XLI. fig. 4.

SPECIES. THECLA ACHÆUS: alis subbicaudatis fuscis; maculis flavis; subtus flavis maculis aureis, numerosis, quibusdam fusco-cinctis. Expans. alar. unc. $1\frac{3}{4}$.

THECLA: with two short tails, wings brown, above with yellow spots; beneath pale yellow, with numerous golden spots, some of which are edged with purplish brown. Expanse of the wings, $1\frac{3}{4}$ inch.

SYN. Hesperia (R.) Achæus, *Fabr. Ent. Syst.* III. 1. p. 273. *Cramer Pap.* pl. 352, f. G, H. *Enc. Méth.* 9. 644, (Polyommatus A.)

HABITAT. "In India," (*Fabricius*). Surinam, (*Enc. Méth.*)

LEPIDOPTERA.

THECLA PHORBAS.

Plate XLI. fig. 5.

SPECIES. THECLA PHORBAS: alis caudatis, fuscis, disco albo; subtus albis, cinereo strigosis, punctis duobus anguli ani atris. Expans. alar. unc. 1⅖.

THECLA: with the wings tailed, brown, with a white disc; beneath white with ash-coloured streaks, and with two small black eyes at the anal angle. Expanse of the wings, 1⅖ inch.

SYN. Hesperia (R.) Phorbas, *Fab. Ent. Syst.* III. 1. *p.* 277. *Enc. Méth.* 9. 646, (Polyommatus P.)

HABITAT. "In India," (*Fabricius*).

THECLA ÆOLUS. ♂

Plate XLII. fig. 1.

SPECIES. THECLA ÆOLUS: alis caudatis, cyaneis, nitidis, margine fusco, macula nigra; subtus fuscis, fascia communi alba nigroque striata. Expans. alar. unc. 1¾.

THECLA: with the wings tailed, above blue in the disc, shining with a brown margin and a black central spot; beneath brown with a common white band, and with black streaks. Expanse of the wings, 1¾ inch.

SYN. Hesperia Æolus, *Fabr. Ent. Syst.* III. 1. *p.* 284. *Enc. Méth* 9. 628, (Polyommatus Æ.)

Papilio Thallus, *Cramer Pap. pl.* 259, *f. C, D.*

♀ Hesperia Pelion, *Fabr. Ent. Syst.* III. 1. *p.* 213. *Cramer, pl.* 6, *f. E, F.*

HABITAT. "In Indiis," (*Fabricius*). Guiana, (*Enc. Méth.*)

THECLA STREPHON.

Plate XLII. fig. 2.

SPECIES. THECLA STREPHON: alis caudatis, fuscis; disco cærulescentibus; subtus cinereis fascia alba, anguloque ani ocello gemino rufo pupilla nigra. Expans. alar. unc. 1½.

THECLA: with the wings tailed, brown; the disc bluish; beneath cinereous with a white fascia, and with a double red ocellus, having a black pupil, at the anal angle. Expanse of the wings, 1½ inch.

SYN. Hesperia, (R.) Strephon, *Fabr. Ent. Syst.* III. 1. *p.* 281. *Enc. Méth.* 9. 632 ? (Polyommatus S.)

Papilio Cyllarus, *Cramer Pap. pl.* 27, *fig. C, D.*

HABITAT. "In India orientali," (*Fabricius*). Surinam, and Brazil, (*Enc. Méth.*)

LEPIDOPTERA.

THECLA PHILIPPUS.

Plate XLII. fig. 3.

SPECIES. THECLA PHILIPPUS: alis caudatis, fuscis; subtus albis; posticis supra striga postica alba punctisque duobus subocellaribus anguli ani atris. Expans. alar. unc. 1¼.

THECLA: with the wings tailed, brown; beneath white; the posterior above with a posterior white streak, and two eyelets at the anal angle of a black colour. Expanse of the wings, 1¼ inch.

SYN. Hesperia R. Philippus, *Fabr. Ent. Syst.* III. 1. p. 283. *Enc. Méth.* 9. 646. (Polyommatus P.)

HABITAT. "In India," (*Fabricius*).

THECLA PERICLES.

Plate XLII. fig. 4.

SPECIES. THECLA PERICLES: alis bicaudatis, nigris, immaculatis; subtus fuscis albo undatis; posticis angulo ani macula duplici argentea, punctisque aliquot nigris. Expans. alar. unc. 1¼.

THECLA: with two tails, wings black, immaculate; beneath brown with white waved lines; posterior with a double silvery spot at the anal angle, and with several black points. Expanse of the wings, 1¼ inch.

SYN. Hesperia (R.) Pericles, *Fabr. Ent. Syst.* III. 1. p. 273. *Enc. Méth.* 9. 622, (Polyommatus P.)

HABITAT. "In Indiis," (*Fabricius*).

ERYCINA THUCYDIDES.

Plate XLIII. fig. 1.

SPECIES. ERYCINA THUCYDIDES: alis integerrimis, nigris; macula magna media fulva in singula; subtus cinereis rufo undatis. Expans. alar. unc. 1½.

ERYCINA: with the wings entire, black, with a large fulvous spot in the centre of each; beneath cinereous with red wavy lines. Expanse of the wings, 1½ inch.

SYN. Hesperia (R.) Thucydides, *Fabr. Ent. Syst.* III. 1. p. 323. *Enc. Méth.* 9. 589. (Erycina T.)

HABITAT. "In Indiis," (*Fabricius*).

LEPIDOPTERA.

ERYCINA PETRONIUS.
Plate XLIII. fig. 2.

SPECIES. ERYCINA PETRONIUS: alis integris cæruleis, strigis apiceque nigris ; subtus fusco-cinereis, nigro punctatis. Expans. alar. unc. 1¾.
ERYCINA : with the wings entire, blue, with black streaks and posterior margins ; beneath brownish ashy, with black spots. Expanse of the wings, 1¾ inch.
SYN. Hesperia (R.) Petronius, *Fabr. Ent. Syst.* III. 1. *p.* 324. *Enc. Méth.* 9. 570. (Erycina P.)
♀ Papilio Menander, *Cramer Pap. tab.* 334, *f. C, D.*
HABITAT. "In Indiis," (*Fabricius*). Guiana and Brazil, (*Enc. Méth.*)

ERYCINA REGULUS.
Plate XLIII. fig. 3.

SPECIES. ERYCINA REGULUS: alis integerrimis, nigris, fasciis duabus albis aut flavescentibus, externa anticarum interrupta. Expans. alar. unc. 1⅓.
ERYCINA : with the wings entire, black, with two white or yellowish broad bands, the outer in the anterior wings being broken. Expanse of the wings, 1⅓ inch.
SYN. Hesperia (R.) Regulus, *Fabr. Ent. Syst.* III. 1. *p.* 318. *Enc. Méth.* 9. 589. (Erycina R.)
HABITAT. "In Indiis," (*Fabricius*). Brazil, (*Enc. Méth.*)

ERYCINA LUCANUS.
Plate XLIII. fig. 4.

SPECIES. ERYCINA LUCANUS: alis integerrimis, supra flavis, macula media anticarum limboque omni nigris ; subtus anticis flavescentibus fusco-maculatis ; posticis rubris flavo irroratis et fusco-maculatis. Expans. alar. unc. 1¼.
ERYCINA : with the wings entire, above yellow, with a central spot in the anterior pair and all the edges black ; beneath the anterior yellowish, with brown spots ; the posterior red, with yellow dots and brown spots. Expanse of the wings, 1¼ inch.
SYN. Hesperia (R.) Lucanus, *Fabr. Syst. Ent.* III. 1. *p.* 322. *Enc. Méth.* 9. 586, (Erycina L.)
HABITAT. "In Indiis," (*Fabricius*).

The accompanying figure of a species described by Fabricius, from the collection of Drury, and the drawings of Mr. Jones, (from which also Donovan, in all probability, obtained his acquaintance with the species,) enables us to clear up the con-

LEPIDOPTERA.

fusion in the Fabrician description, in which the wings are described—"*supra* flavæ macula media anticarum, *disco* que omni nigris"; the description in the short specific character being "alis nigris, *disco* flavo." Hence it is evident that, in the former description, the term "disco" was accidentally used instead of "limbo." In the Encyclopédie Méthodique the confusion is increased, by the employment of the word "dessous," instead of "dessus."*

ERYCINA TARQUINIUS.

Plate XLIV. fig. 1.

SPECIES. ERYCINA TARQUINIUS: alis integris, nigris; anticis macula oblonga baseos sinuata postice triloba; posticis angulo anali late flavo punctis quinque nigris; subtus cinereis maculis nigris et rufescentibus, his albo cinctis. Expans. alar. unc. 1⅓.

ERYCINA: with the wings entire, black; the anterior with an oblong sinuated spot at the base, the extremity of which is trilobed; the posterior with a broad yellow patch at the anal angle, having five black dots; beneath cinereous, with black and red spots, the latter encircled with white. Expanse of the wings, 1⅓ inch.

SYN. Hesperia (R.) Tarquinius, *Fabr. Ent. Syst.* III. 1. p. 319. *Enc. Méth.* 9. 580, (Erycina T.)

HABITAT. "In Indiis," (*Fabricius*).

ERYCINA ÆMULIUS.

Plate XLIV. fig. 2.

SPECIES. ERYCINA ÆMULIUS: alis integris, cinereis, lineolis transversis albidis maculisque oblongis fuscis; posticis supra pallidis; subtus omnibus maculis plurimis fuscis. Expans. alar. unc. 1¼.

ERYCINA: with the wings entire, cinereous, with short transverse whitish lines and oblong brown spots; the posterior pair above pale; the under surface of all the wings much spotted with brown. Expanse of the wings, 1¼ inch.

SYN. Hesperia (R.) Æmulius, *Fabr. Ent. Syst.* III. 1. p. 322. *Enc. Méth.* 9. 580, (Erycina Æ.)

---

* In the first edition of this work, Erycina Salustius was introduced in the text, after the foregoing species; but no figure of it appeared in the plates, I have therefore omitted it.—(J. O. W.)

LEPIDOPTERA.

HESPERIA NUMITOR.

Plate XLIV. fig. 3.

FAMILY. HESPERIIDÆ, *Stephens*.
GENUS. HESPERIA, *Latreille, Godart*. Battus p. *Scop.* Papilio (Pl. Rur.) *Linnæus*.
SPECIES. HESPERIA NUMITOR: alis integerrimis, fuscis; posticis supra disco flavo; subtus flavis, anticis disco fusco. Expans. alar. unc. 1—1¼.
    HESPERIA: with the wings entire, brown; the posterior with the disc above yellow; the under surface of all the wings yellow, except the disc of the anterior, which is brown. Expanse of the wings, 1 to 1¼ inch.
SYN. Hesperia (R.) Numitor, *Fabr. Ent. Syst.* III. 1. *p.* 324. *Enc. Méth.* 9. 777.
HABITAT. "In Indiis," (*Fabricius*). Philadelphia, (*Enc. Méth.*)

Donovan has evidently introduced this, as well as many of the preceding butterflies, (especially those belonging to the great genus Erycina, which are almost exclusively Brazilian,) into the present work, on the authority of the loose habitat of Fabricius, "In Indiis." The Hesperia Numitor is, however, an inhabitant of North America, having been sent from Philadelphia to M. Latreille by M. Lesueur. I have also received it from the same city, in a large collection of insects of that country, for which I am indebted to the liberality of Mr. Titian R. Peale, the curator of the Museum of Philadelphia, a very zealous entomologist.

POLYOMMATUS PLINIUS.

Plate XLV. fig. 1.

SPECIES. POLYOMMATUS PLINIUS: alis caudatis, albo nigroque variis; anticis supra maculis quadratis fuscis; omnibus subtus albis, fusco et nigro transverse strigosis, serie duplici punctorum submarginalium punctoque gemmo aurea ad angulum ani posticarum. Expans. alar. unc. 1.
    POLYOMMATUS: with the wings tailed, and varied above with black and white; the anterior above with square brown spots; all the wings beneath white, with transverse brown or black streaks, and with a double row of submarginal dots; the posterior with a double golden spot at the anal angle. Expanse of the wings, 1 inch.
SYN. Hesperia (R.) Plinius, *Fabr. Ent. Syst.* III. 1. *p.* 284. *Enc. Méth.* 9. 658, (Polyommatus Pl.) *Horsfield Lep. Jav.* 1. *p.* 72.

LEPIDOPTERA.

POLYOMMATUS PLATO.

Plate XLV. fig. 2.

SPECIES. POLYOMMATUS PLATO: alis caudatis, supra cæruleis limbo fusco; posticis punctis marginalibus atris; subtus cinereis albo undatis, posticis ocello atro iride flava, pupillaque duplici argentea et ad angulum ani macula argentea. Expans. alar. unc. 1¼.

POLYOMMATUS: with the wings tailed; above blue with a brown border; the posterior with marginal black dots; beneath cinereous with white waves, the posterior with a black ocellus, having the iris yellow, and a double silvery pupil, a silvery spot also at the anal angle. Expanse of the wings, 1¼ inch.

SYN. Hesperia (R.) Plato, *Fabr. Ent. Syst.* III. 1. *p.* 288?
Polyommatus Plato, *Enc. Méth.* 9. 655?

HABITAT. "In Indiis," (*Fabricius*).

The description of the ocelli on the under side of the under wings of the insect described under this name, not agreeing with the Fabrician description, I have given the quotation with a mark of doubt.

POLYOMMATUS HIPPOCRATES.

Plate XLV. fig. 3.

SPECIES. POLYOMMATUS HIPPOCRATES: alis caudatis; supra fuscis, anticis apice, posticis margine tenui albis; subtus albidis, anticis striga punctorum, posticis punctis sparsis, nigris. Expans. alar. unc. 1.

POLYOMMATUS: with the wings tailed; above brown, the tips of the anterior and a slender margin to the posterior white; beneath whitish, the anterior with a row of black spots, and the latter with several irregularly placed black spots. Expanse of the wings, 1 inch.

SYN. Hesperia (R.) Hippocrates, *Fabr. Ent. Syst.* III. 1. *p.* 288. *Enc. Méth.* 9. 659, (Polyommatus H.)

THECLA THEOCRITUS.

Plate XLV. fig. 4.

SPECIES. THECLA THEOCRITUS: alis caudatis; supra virescentibus aureo nitidulis, costa lata obscuriore; subtus nigris, strigis aliquot e punctis parvis flavescentibus. Expans. alar. unc. 1½.

LEPIDOPTERA.

THECLA: with the wings tailed; above greenish with golden spangles, and with the front margin and tip dark; beneath black, with several rows of small yellowish dots. Expanse of the wings, 1½ inch.

SYN. Hesperia (R.) Theocritus, *Fabr. Ent. Syst.* III. 1. *p.* 289. *Enc. Méth.* 9. 653, (Polyommatus Th.)

POLYOMMATUS PARRHASIUS.

Plate XLV. fig. 5.

SPECIES. POLYOMMATUS PARRHASIUS: alis caudatis; supra cæruleis margine fusco, (♂), aut fuscis, (♀); posticis ante marginem striga punctorum nigrorum ocellatorum; omnibus subtus cinereis, pone medium albo undatis; posticis punctis tribus baseos atris, albo cinctis, apiceque punctis quatuor aureis, tertio puncto atro. Expans. alar. ¾—1 unc.

POLYOMMATUS: with the wings tailed; above blue with a brown margin in the male, brown in the female, the posterior with a submarginal row of black ocellated spots; the under side of all the wings cinereous, with white waves beyond the middle, the posterior with three black spots, edged with white towards the base, and with four golden marginal spots, the third having a black dot in the middle. Expanse of the wings, ¾ to 1 inch.

SYN. Hesperia (R.) Parrhasius, *Fabr. Ent. Syst.* III. 1. *p.* 289. *Enc. Méth.* 9. 657, (Polyommatus P.) *Horsfield Lep. Jav.* 1. *p.* 86 ?

HABITAT. "In India," (*Fabricius, Enc. Méth.*) Java, (*Horsfield*).

As in our little English "blues," (as the species of Polyommatus are called by collectors,) there are many species so very closely allied together, that it is very difficult to settle their specific distinctions, so there appear to be several Indian species so nearly related to Pol. Parrhasius, that it is only by precisely following the particular description of that species given by Fabricius, (and which I have embodied in the above definition,) that we can avoid confusion. The species was known to Fabricius, through the drawings of Mr. Jones, and the collection of Mr. Drury, to both of which Donovan had access, so that we may adopt his figures as the real representatives of the Fabrician insect. It is on these grounds that I have attached a mark of doubt to the reference to Dr. Horsfield's Lepidoptera Javanica. Donovan has, however, fallen into the strange mistake of considering the larger individual with blue wings as the female, and the smaller brown one as the male, whereas, as in the common English "blues," the opposite is the case.

LEPIDOPTERA.

ERYCINA? BIBULUS.

Plate XLVI. fig. 1.

SPECIES. ERYCINA? BIBULUS: alis integerrimis; supra fuscis, anticis in medio litura cinerea; subtus anticis cinereis punctis aliquot apicis fuscis annulo argenteo cinctis, strigaque marginali argentea; posticis albis striga media argentea, punctis aliquot fuscis apice fusco punctis argenteis. Expans. alar. unc. $1\frac{1}{4}$.

ERYCINA? with the wings entire; above brown, the anterior with a cinereous stripe in the centre; beneath the anterior cinereous, with several apical brown spots surrounded with silver, and a marginal silver stripe; posterior white, with a central silver bar, several brown spots, and the apex brown with silvery spots. Expanse of the wings, $1\frac{1}{4}$ inch.

SYN. Hesperia (R.) Bibulus, *Fabr. Ent. Syst.* III. 1. *p.* 307.

HABITAT. "In Indiis," (*Fabricius*).

POLYOMMATUS HYLAX.

Plate XLVI. fig. 2.

SPECIES. POLYOMMATUS HYLAX: alis integris, supra fuscis immaculatis, subtus cinereis, punctis nigris arcubus submarginalibus fuscis. Expans. alar. $\frac{5}{8}$ unc.

POLYOMMATUS: with the wings entire, brown above without spots, beneath cinereous with black spots, and with a row of submarginal lunules, (behind which is a row of small black dots). Expanse of the wings, $\frac{5}{8}$ of an inch.

SYN. Hesperia (R.) Hylax, *Fabr. Mant. Ins.* 2. *p.* 77. *Ent. Syst.* III. 1. *p.* 304. *Enc. Méth.* 9. 701, (Polyommatus H.) *nec* Pithecops Hylax, *Horsfield Lep. Jav.* 1. *p.* 66, *pl.* 1. *f.* 2, 2 *a*.

Papilio Lysimon, *Ochsenheimer Pap. Eur. t.* 1? *Enc. Méth.* 9. 701?

HABITAT. "In India orientali," (*Fabricius*).

I can by no means agree with Dr. Horsfield, in considering the insect which he has figured under the name of Pithecops Hylax as identical with the Fabrician insect, as that which is here figured is marked more agreeably with the original description of the species, short as it is. Had Fabricius been describing Dr. Horsfield's insect, he would have said "strigis duabus submarginalibus punctorum nigrorum," instead of "nigropunctatis," which implies a general scattering of dots over the wings; moreover Dr. Horsfield's insect exhibits no trace of the "arcubus fuscis," described by Fabricius.

LEPIDOPTERA.

ERYCINA ATHEMON. ♂.

Plate XLVI. fig. 3.

SPECIES. ERYCINA ATHEMON: alis integris, albis; anticis margine exteriori nigro, cæruleo alboque maculato. Expans. alar. 1—1½ unc.

ERYCINA: with the wings entire, white; the anterior having the front margin broadly marked with black, in which are several blue and white spots. Expanse of the wings, 1 to 1½ inch.

SYN. Papilio Athemon, *Linn. Syst. Nat.* 2. 792. *Fabr. Ent. Syst.* III. 1. *p.* 318, (♀).
Hesperia (R.) Cœnus, *Fabr. Ent. Syst.* III. 1. p. 308. (♂). *Donovan, 1st edit.*
Erycina Arthemon, *Enc. Méth.* 9. 578.

HABITAT. " In Indiis," (*Fabricius*, Cœnus). In America, (*Linnæus*, Athemon). Brazil, (*Enc. Méth.*)

The name of Athemon having been given to the female of this species, I have adopted it in preference to following Donovan and using that of Cœnus, proposed by Fabricius for the male, rejecting the latter name in consequence of its similarity to Ceneus, a name employed by Fabricius for another species of Erycina. It is customary, however, in such cases, to adopt the specific name of the male rather than of the female, where they happen to have been previously described under different names.

ERYCINA? LIVIUS.

Plate XLVI. fig. 4.

SPECIES. ERYCINA? LIVIUS: alis integerrimis; supra atris, macula disci cærulea; subtus cinereis; anticis macula oblonga baseos flava, argenteo marginata; posticis fasciis tribus rufis argenteo marginatis, margine flavo, fascia argentea. Expans. alar. 1⅙ unc.

ERYCINA? with the wings entire; above black, with a discoidal blue spot; beneath cinereous; the anterior with an oblong basal yellow spot, margined with silver; the apex yellow, with a broad brown bar, margined with silver; the posterior with three red bars margined with silver, the margins yellow with a silver line. Expanse of the wings, 1⅙ inch.

SYN. Hesperia (R.) Livius, *Fabr. Ent. Syst.* III. 1. p. 315. *Enc. Méth.* 9. 825, (Nymphalis L.)

HABITAT. " In Indiis," (*Fabricius*).

LEPIDOPTERA.

THECLA ROMULUS.

Plate XLVI. fig. 5.

SPECIES. THECLA ROMULUS: alis integerrimis; supra fuscis, immaculatis; subtus viridibus, anticis immaculatis, posticis striga fusca, maculaque postica rufa. Expans. alar. unc. 1⅙.

THECLA: with the wings entire; above brown, immaculate; beneath green, the anterior without spots, the posterior with a brown stripe, and a submarginal red ocellus. Expanse of the wings, 1⅙ inch.

SYN. Hesperia (R.) Romulus, *Fabr. Ent. Syst.* III. 1. p. 316. *Enc. Méth.* 9. 674. (Polyommatus R.)

HABITAT. "In Indiis," (*Fabricius*).

The accompanying figure exhibits no trace of the brown line on the under surface of the posterior wings described by Fabricius, from the drawings of Mr. Jones, and the collection of Drury.

ERYCINA PTOLOMÆUS. ♂.

Plate XLVI. fig. 6.

SPECIES. ERYCINA PTOLOMÆUS: alis integerrimis; supra fereugineis, strigis plurimis fuscis, ultima ante marginem punctata; subtus (mas) cæruleis, (fem.) rufis, basi cærulescente. Expans. alar. unc. 1½.

ERYCINA: with the wings entire; above ferruginous with several brown streaks, the submarginal one being composed of points; beneath in the male blue, in the female ferruginous, with the base paler blue. Expanse of the wings, 1½ inch.

SYN. Hesperia (R.) Ptolomæus, *Fabr. Ent. Syst.* III. p. 319, ♂. *Enc. Méth.* 9. 572, (Erycina Pt.)

Hesperia (R.) Lucius, *Fabr. Ent. Syst.* III. 1. p. 320, ♀.

HABITAT. "In Indiis," (*Fabricius*). Brazil, (*Enc. Méth.*).

ERYCINA OVIDIUS.

Plate XLVI. fig. 7.

SPECIES. ERYCINA OVIDIUS: alis integris, supra fulvis, subtus flavescentibus; utrinque strigis sex undatis, feminæ nigris, maris aureo pulverulentis. Expans. alar. unc. 1¼.

LEPIDOPTERA.

ERYCINA: with the wings entire, above fulvous, beneath yellowish, on each side with six waved stripes, black in the female, but spotted with gold in the male. Expanse of the wings, 1¼ inch.

SYN. Hesperia (R.) Ovidius, *Fabr. Ent. Syst.* III. 1. p. 320. *Ent. Méth.* 9. 571.
Papilio Fatima, *Cramer, pl.* 271, *f. A, B, ♂. C, D, ♀*.

HABITAT. "In Indiis." (*Fabricius*). East Indies, (*Enc. Méth.*)

The seven preceding species are represented upon the Adiantum succulentum.

LEPTOCIRCUS CURIUS.

Plate XLVII. fig. 1.

FAMILY. PAPILIONIDÆ.
GENUS. LEPTOCIRCUS, *Swainson.* (Lamproptera, *G. R. Gray.* Erycina p. *Enc. Méth.* Iphiclides, *Hubner.*)
SPECIES. LEPTOCIRCUS CURIUS: alis concoloribus, nigris, fascia communi glauca; anticis ante apicem hyalinis. Expans. alar. unc. 1¾.
LEPTOCIRCUS: with the wings concolorous, black, with a common greenish white fascia; the anterior having a large transparent spot near the apex. Expanse of the wings, 1¾ inches.

SYN. Papilio (Eq. Ach.) Curius, *Fabr. Ent. Syst.* III. 1. p. 28. *Enc. Méth.* 9. 564, 827, (Erycina C.)
Leilus [Urania] (Leptocircus) Curius, *Swainson Zool. Illust.* 2d. series, *pl.* 106.
Leptocircus Curius, *Boisduval Hist. Nat. Lepid.* 1. 381, *pl.* 17, *f.* 3. *Naturalist's Library Entomol.* vol. 5, *pl.* 5, *fig.* 1.
Erycina (Lamproptera) Curius, *G. R. Gray in Griff. An. Kingd. Ins. pl.* 102, *f.* 4.

HABITAT. Siam, (*Fabricius*). Java, (*Boisduval*).

This insect is exceedingly interesting, with reference to the general classification of the diurnal butterflies. It was, by Fabricius, regarded as a true Papilio, and certainly the head, palpi, antennæ, legs, &c. clearly prove the correctness of this affinity. In the Encyclopédie Méthodique, however, it was considered as an Erycina, doubtless from its small size, and its apparent resemblance to some of the long-tailed species, such as Rhetus and Dorilas of Cramer. Donovan also asserted its greater affinity to the Papiliones Plebeii. Mr. Swainson next proved it to belong to the family Papilionidæ, by his figures of the various parts in detail; but, from too eagerly

LEPIDOPTERA.

following out his peculiar theories, he gave it as a subgenus of the genus Urania, which name he changed to Leilus, (considering its type to be the Papilio Leilus of Linnæus,) on the ground that the name Urania had been employed in Botany, which is certainly the case; but as the genus of plants ought to retain its *previous* name Ravenala, the name of Urania is not inapplicable to a genus of Lepidoptera. But Leptocircus ought clearly not to be regarded as belonging to the same group as Leilus, Orontes, and Rhipheus, together with Patroclus, Lunus, and Empedocles, which decidedly belong thereto, although omitted by Mr. Swainson. The veins of the wings of Leptocircus are decidedly Papilionideous, and have no resemblance whatever to those of Urania. There is it is true some analogy in the *form* of the posterior wings, but, I presume, no real relation can be supported on such a character, and I can discover no other to give it a relation either with Urania, or Erycina. In the male of this insect, the pale band crossing the wings is of a greenish white, but in the female it is transparent white in the anterior, and white in the posterior wings.

The species was described by Fabricius from a specimen in the collection of Sir J. Banks. It is become sufficiently common.

HESPERIA PROPERTIUS.
Plate XLVII. fig. 2.

FAMILY. HESPERIIDÆ.
GENUS. HESPERIA.
SPECIES. HESPERIA PROPERTIUS: alis supra nigris, anticis lineolis duabus marginalibus baseos maculisque flavis, posticis fascia flava; subtus, anticis atro, flavo, rufoque variis, posticis fasciis rufis flavisque margine tenuiori nigro. Expans. alar. unc. 1½.
HESPERIA: with the wings above black, the anterior with two marginal lines at the base and some spots of yellow, the posterior with a yellow bar; beneath the anterior are varied with black, yellow and red, and the posterior have alternate bars of red and yellow, with a slender black edge. Expanse of the wings, 1½ inch.
SYN. Hesperia (U.) Propertius, *Fabr. Ent. Syst.* III. 1. p. 325. *Enc. Méth.* 9. 773.
HABITAT. "In India," (*Fabricius*).

HESPERIA TIBULLUS.
Plate XLVII. fig. 3.

SPECIES. HESPERIA TIBULLUS: alis supra anticis nigris flavo-maculatis; subtus subconcoloribus, costa subrufa; posticis flavis margine omni, postico interrupto, nigro, subtus variegatis. Expans. alar. unc. 1½.

LEPIDOPTERA.

HESPERIA: with the anterior wings above black, spotted with yellow; beneath similarly coloured, but with a reddish costa; posterior pair yellow with the margin (interrupted behind) black, beneath variegated. Expanse of the wings, 1½ inch.

SYN. Hesperia (U.) Tibullus, *Fabr. Ent. Syst.* III. 1. *p.* 326. *Enc. Méth.* 9. 754.

HABITAT. "In Indiis," (*Fabricius*).

HESPERIA AUGIAS.

Plate XLVIII. fig. 1.

SPECIES. HESPERIA AUGIAS: alis fulvis; supra anticis linea obliqua, limboque postico nigris; posticis nigris, puncto subcentrali fasciaque postica utrinque dentatis rufo-flavis. Expans. alar. unc. 1¼.

HESPERIA: with fulvous wings; above the anterior are marked by an oblique line, and the posterior edge of black; the posterior are black, with a central spot and a posterior bar, toothed on each side, of reddish yellow. Expanse of the wings, 1¼ inch.

SYN. Papilio (Pleb. Urb.) Augias, *Linn. Amœn. Acad. t.* 4, *p.* 410. *Syst. Nat.* 2. 794. *Fabr. Ent. Syst.* III. 1. *p.* 327. *Enc. Méth.* 9. 767, (Hesperia A.)

HABITAT. Java, (*Linnæus Enc. Méth.*) India, (*Fabricius*).

The specimen here figured appears to be less strongly marked with black than ordinary, the posterior margin of the wings being spotted with that colour, instead of being entirely black.

HESPERIA ORIGINES.

Plate XLVIII. fig. 2.

SPECIES. HESPERIA ORIGINES: alis fuscis; striga punctorum alborum; anticis basi obliqué testaceis. Expans. alar. unc. 1¼.

HESPERIA: with the wings brown, with a row of white dots; the anterior with a large oblique testaceous patch at the base. Expanse of the wings, 1¼ inch.

SYN. Hesperia (U.) Origines, *Fabr. Ent. Syst.* III. 1. *p.* 328, (Hesperia Thaumas). *Enc. Méth.* 9. 766?

HABITAT. "In Indiis," (*Fabricius.*) North America, (*Enc. Méth.*)

As there appears to be some doubt of the precise identity of this insect with the Hesperia Thaumus of Fabricius, as well as of its sex, and as the accompanying figure agrees with the Fabrician description of H. Origines, I have thought it best to retain the latter name.

LEPIDOPTERA.

ERYCINA ? PLUTARGUS.

Plate XLVII. fig. 3.

SPECIES. ERYCINA ? PLUTARGUS : alis supra fuscis auro irroratis, anticis macula apicali testacea; subtus pallidioribus, anticis margine exteriori testaceo, posticis strigis punctoque obscurioribus. Expans. alar. unc. 1⅛.

ERYCINA? with the wings above brown with gold spangles, and with a testaceous spot at the tips of the anterior pair; beneath paler, the anterior with the front margin testaceous; the posterior pair with streaks and a spot of darker colour. Expanse of the wings, 1⅛ inch.

SYN. Hesperia (U.) Plutargus, *Fab. Ent. Syst.* III. 1. *p.* 329. *Enc. Méth.* 9. 776 (Hesperia P.)

HABITAT. "In Indiis," (*Fabricius*).

HESPERIA EPICTETUS.

Plate XLVIII. fig. 4.

SPECIES. HESPERIA EPICTETUS: alis fuscis, disco flavo; anticarum macula longitudinali fusca et in hac lunula flava; posticis subtus flavis immaculatis. Expans. alar. unc. 1.

HESPERIA: with the wings brown, the disc being yellow; the anterior pair having a longitudinal brown spot running through the middle of the wing, in which is a yellow lunule; the posterior pair are yellow and without spots beneath. Expanse of the wings, 1 inch.

SYN. Hesperia (U.) Epictetus, *Fabr. Ent. Syst.* III. 1. *p.* 330. *Enc. Méth.* 9. 768.

HABITAT. "In Indiis," (*Fabricius*). Brazil, (*Enc. Méth.*)

HESPERIA CHEMNIS.

Plate XLIX. fig. 1.

SPECIES. HESPERIA CHEMNIS : alis subcaudatis, fuscis basi magis brunneis; anticis maculis sex, posticis tribus quadratis hyalinis, margine postico flavo. Expans. alar. unc. 2½.

HESPERIA: with the wings slightly tailed, brown, with the base more fulvous; anterior pair with six, posterior with three square hyaline spots; posterior margin yellow. Expanse of the wings, 2½ inches.

SYN. Hesperia (U.) Chemnis, *Fabr. Ent. Syst.* III. 1. *p.* 331. *Enc. Méth.* 9. 746.

HABITAT. "In Indiis," (*Fabricius*).

LEPIDOPTERA.

HESPERIA THRAX.

Plate XLIX. fig. 2.

SPECIES. HESPERIA THRAX: alis ecaudatis, fuscis, maculis tribus fenestratis, exteriore minore. Expans. alar. unc. 2½.

HESPERIA: without tails, with brown wings, the anterior having three fenestrated spots, of which the outer one is smallest. Expanse of the wings, 2½ inches.

SYN. Papilio (Pl. Urb.) Thrax, *Linn. Syst. Nat.* 2. 794. *Enc. Méth.* 9. 748, (Hesperia T.) *Fabr. Ent. Syst.* III. 1. p. 337 ?

HABITAT. Java, (*Linnæus, Enc. Méth.*)

Donovan expressly states that his specimen of this insect corresponded with that in the Linnæan cabinet. It also agrees with the description of a Javanese butterfly described in the Encyclopédie Méthodique as the true Thrax. This circumstance renders doubtful the references to Fabricius, Clerk, and Cramer, who has figured three species, regarded as varieties of Thrax, under the names of Papilio Salus, Sebaldus, and Ramasis.

HESPERIA (EANTIS) MITHRIDATES.

Plate XLIX. fig. 3.

SUBGENUS. EANTIS, *Boisduval.*

SPECIES. HESPERIA (EANTIS) MITHRIDATES: alis atris, purpureo-maculatis, fasciaque submarginali purpurea lunulis albidis; subtus nigris, fasciis duabus macularibus purpureis. Expans. alar. unc. 2.

HESPERIA (EANTIS): with the wings black, spotted with purple, and with a purple submarginal fascia, in which are whitish lunules; beneath black, with two rows of purple spots. Expanse of the wings, 2 inches.

SYN. Hesperia (U.) Mithridates, *Fabr. Ent. Syst.* III. 1. p. 336. *Enc. Méth.* 9. 792, (Hesperia M.)

HABITAT. "In Indiis," (*Fabricius*). Brazil ? (*Enc. Méth.*)

LEPIDOPTERA.

HESPERIA (EANTIS) THRASIBULUS.

Plate XLIX. fig. 4.

SPECIES. HESPERIA (EANTIS) THRASIBULUS : alis atris cæruleo undulatis ; posticis subtus angulo ani cinereo fusco punctato. Expans. alar. unc. 1½.

HESPERIA (EANTIS) : with black wings, undulated with blue ; the posterior beneath with the anal angle cinereous, spotted with brown. Expanse of the wings, 1½ inch.

SYN. Hesperia (U.) Thrasibulus, *Fabr. Ent. Syst.* III. 1. p. 346. *Enc. Méth.* 9. 792.

HABITAT. "In Indiis," (*Fabricius.*) Brazil, (*Enc. Méth.*)

HESPERIA JOVIANUS.

Plate L. fig. 1.

SPECIES. HESPERIA JOVIANUS : alis atris, anticis macula transversa pone medium (alteraque minori subapicali in ♀) alba, strigaque marginali punctorum oblongorum cæruleorum ; posticis cæruleo striatis et in his maculæ aliquot oblongæ albæ. Expans. alar. unc. 1½.

HESPERIA : with the wings black ; the anterior with a transverse bar of white beyond the middle, (a smaller subapical one in the female,) and a submarginal row of oblong blue dots ; the posterior pair with blue streaks, in which are several oblong white spots. Expanse of the wings, 1½ inch.

SYN. Hesperia (U.) Jovianus, *Fabricius Ent. Syst.* III. 1, p. 348. *Enc. Méth.* 9. 788.

HABITAT. "In Indiis," (*Fabricius*). Brazil, (*Enc. Méth.*)

HESPERIA SALVIANUS.

Plate L. fig. 2.

SPECIES. HESPERIA SALVIANUS : alis supra fuscis, viridi maculatis strigaque postica macularum obscuriorum ; subtus anticis fuscis fascia lata alba, posticis albis striga marginali e punctis septem fuscis. Expans. alar. unc. 1½.

HESPERIA : with the upper side of the wings brown, mottled with green, and with a submarginal row of obscure dots ; beneath the anterior wings brown with a broad white bar, posterior white with a marginal row of seven brown dots. Expanse of the wings, 1½ inch.

SYN. Hesperia (U) Salvianus, *Fabr. Ent. Syst.* III. 1. p. 348. *Enc. Méth.* 9. 789.

HABITAT. "In Indiis," (*Fabricius*).

LEPIDOPTERA.

HESPERIUS GALENUS.

Plate L. fig. 3.

SPECIES. HESPERIA GALENUS: alis fuscis punctis maculisque flavis; anticis fascia maculari transversa media; posticis macula magna oblonga e medio ad marginem extensa. Expans. alar. unc. 1½.

HESPERIA: with the wings brown with yellow dots and spots; the anterior with a central transverse row of spots forming a bar, and the latter with a large oblong spot reaching from the middle to the tips. Expanse of the wings, 1½ inch.

SYN. Hesperia (U.) Galenus, *Fabr. Ent. Syst.* III. 1, p. 350. *Enc. Méth.* 9. 773.

HABITAT. "In Indiis," (*Fabricius*).

HESPERIA CATULLUS.

Plate L. fig. 4.

SPECIES. HESPERIA CATULLUS: alis atris, striga communi submarginali punctorum alborum; anticis etiam punctis nonnullis albis, subtus paucioribus. Expans. alar. unc. 1⅛.

HESPERIA: with the wings black, with a row of submarginal white dots on all the wings; the anterior pair have also a few white dots scattered over the surface, being fewer beneath. Expanse of the wings, 1⅙ inch.

SYN. Hesperia (U.) Catullus, *Fabr. Ent. Syst.* III. 1, p. 348. *Abbot and Smith Lep. Georgia*, 1. t. 24. *Enc. Méth.* 9. 777.

HABITAT. "In Indiis," (*Fabricius*). North America, (*Abbot and Smith*).

According to the Encyclopédie Méthodique, this North American insect is liable to considerable variation in the number of the white dots on the upper surfaces of the wings. The caterpillar, according to Abbot and Smith, feeds upon the Monarda punctata. It is of a dark green colour, with two black spots on the first segment; the head is black, with two white dots at the sides.

HESPERIA SPIO.

Plate L. fig. 5.

SPECIES. HESPERIA SPIO: alis reversis, nigris maculis albis plurimis difformibus undique dispersis, subtus simillimis at fuscis. "Similis P. Tages at dimidio minor et brevis."—*Linn.*

HESPERIA: with the wings reversed, black, with many irregular shaped white spots

LEPIDOPTERA.

scattered over the wings; beneath exactly alike, but the black changed to brown. Similar to P. Tages, but smaller by half, and short.

SYN. Papilio (P. N.), Spio, *Linn. Mus. Ulr.* p. 338. (nec 330). *Syst. Nat.* 2. 796. *Fabr. Ent. Syst.* III. 1, *p.* 354.

HABITAT. Cape of Good Hope, (*Linnæus*). India, (*Fabricius*).

Linnæus, both in the Museum Reginæ Ulricæ, and the Systema Naturæ, states that he received this from the Cape of Good Hope, from Tulbagh, who was governor there. I know not upon what authority Fabricius gave it as a native of India.

The five preceding species are represented on the Clerodendrum infortunatum, a plant introduced from the East Indies.

HESPERIA ENNIUS.
Plate LI. fig. 1.

SPECIES. HESPERIA ENNIUS: alis anticis fuscis, maculis hyalinis; posticis supra atris, macula magna media, maculisque marginalibus flavis; subtus fuscis nigro maculatis discoque albo. Expans. alar. unc. $2\frac{1}{10}$.

HESPERIA: with the anterior wings brown, with hyaline spots; the posterior above black, with a large central spot (extending to the base,) and marginal spots of yellow; beneath brown, spotted with black, and with the disc white. Expanse of the wings, $2\frac{1}{10}$ inches.

SYN. Hesperia (U.) Ennius, *Fabr. Ent. Syst.* III. 1. *p.* 338. *Enc. Méth.* 9. 749.

HABITAT. "In Indiis," (*Fabricius*).

HESPERIA POLYBIUS.
Plate LI. fig. 2.

SPECIES. HESPERIA POLYBIUS: alis concoloribus; anticis atris macula media transversa fulva, margine postico niveo; posticis atris, margine postico niveo, anguloque ani fulvo. Expans. alar. unc. $2\frac{1}{4}$.

HESPERIA: with the wings coloured alike on both sides; anterior pair black, with a transverse central fulvous spot, posterior margin white; posterior wings with the margin white and anal angle fulvous. Expanse of the wings, $2\frac{1}{4}$ inches.

SYN. Hesperia (U.) Polybius, *Fabr. Ent. Syst.* III. 1. *p.* 337. *Enc. Méth.* 9. 732.
Hesperia (U.) Palemon, *Fabr. Ent. Syst.* III. 1, *p.* 335. *Sp. Ins.* 2. 134. Cramer, *pl.* 131. *F.*

HABITAT. "In Indiis," (*Fabricius*.) Surinam, Brazil, (*Enc. Méth.*)

LEPIDOPTERA.

HESPERIA ZELEUCUS.

Plate LI. fig. 3.

SPECIES. HESPERIA ZELEUCUS: atra, capite anoque sanguineis; alis concoloribus atris; posticis margine tenuissimo albo. Expans. alar. unc. 2¼.

HESPERIA: black, with the head and anus sanguineous; wings coloured alike on both sides, black; the posterior with a very slender white margin. Expanse of the wings, 2¼ inches.

SYN. Hesperia (U.) Zeleucus, *Fabr. Ent. Syst.* III. 1. *p.* 346. *Enc. Méth.* 9. 733.
Papilio Thasus, *Cramer, pl.* 380, *f. M, N.*

HABITAT. " In Indiis," (*Fabricius*). Surinam, and Brazil, (*Enc. Méth.*)

The three preceding species are represented on a sprig of the Lagerstrœmeria Indica.

HESPERIA ORCUS.

Plate LII. fig. 1.

SPECIES. HESPERIA ORCUS: alis subtus fuscis, anticis punctis maculaque lunata vitreo-flavis; posticis fasciis obsoletis cæruleis: subtus cærulescentibus, margine nigro. Expans. alar. unc. 1½.

HESPERIA: with the wings above brown, the anterior with several spots, and a lunate mark of glassy yellow, the posterior with obsolete blue bars; beneath blue with a black edge. Expanse of the wings, 1½ inch.

SYN. Hesperia (U.) Orcus, *Fabr. Ent. Syst.* III. 1. *p.* 341. *Enc. Méth.* 9. 789.
Papilio Cerialis, *Cramer Pap. pl.* 392, *fig. N. O.* *Stoll Suppl. Cramer, pl.* 10, *fig.* 1.

HABITAT. " In Indiis," (*Fabricius*). Guiana, (*Enc. Méth. Stoll.*)

The caterpillar, according to Stoll, (who observed the transformations of this species in Guiana,) is dirty green, with the head brown, a spot on each side of the segments of the body, and two ventral lines of white. It remains in the chrysalis state only about seven days.

M

LEPIDOPTERA.

HESPERIA BUSIRIS.

Plate LII. fig. 2.

SPECIES. HESPERIA BUSIRIS: thorace albo punctato; alis anticis oblongis, atris, maculis duabus, et inter has punctis duobus flavis, apice cinerascente; posticis fulvis margine atro. Expans. alar. unc. 2.

HESPERIA: with the thorax spotted with white; the anterior wings oblong, black, with two large spots, between which are two dots of yellow, the tip being cinereous; the posterior wings fulvous, with a black margin. Expanse of the wings, 2 inches.

SYN. Hesperia (U.) Busiris, *Fabr. Ent. Syst.* III. 1. *p.* 345. *Enc. Méth.* 9. 758.

HABITAT. "In India," (*Fabricius*).

HESPERIA CELSUS.

Plate LII. fig. 3.

SPECIES. HESPERIA CELSUS: alis supra nigris, subtus brunneo-fuscis; anticis utrinque fascia transversa rufo-flavida; posticis supra immaculatis, infra cinerascenti-pulverulentis, margine externo strigaque transversa brunneo-fuscis. Expans. alar. unc. $1\frac{3}{4}$.

HESPERIA: with the wings above black, beneath rich brown; the anterior on both sides with a transverse fascia of reddish yellow; the posterior pair immaculate above, beneath sprinkled ashy, with the outer margin and a transverse bar rich brown. Expanse of the wings, $1\frac{3}{4}$ inch.

SYN. Hesperia (U.) Celsus, *Fabr. Ent. Syst.* III. 1. *p.* 346. *Enc. Méth.* 9. 759.
Papilio Hiarbas, *Cramer, pl.* 18, *fig. F*?
Hesperia (U.) Thyrsis, *Fabr. Ent. Syst.* III. 1. *p.* 333?

HABITAT. "In Indiis," (*Fabricius.*) Brazil, (*Enc. Méth.*)

ZEUZERA MINEA.

Plate LIII. fig. 1, male. 1 *a*, female.

SECTION. LEPIDOPTERA Nocturna. Phalæna, *Linnæus.*
FAMILY. HEPIALIDÆ, *Leach.*
GENUS. ZEUZERA, *Latreille, Stephens.*
SPECIES. ZEUZERA MINEA: cyanea, alis concoloribus aurantiis, maculis fasciaque media longitudinali cyaneis. Expans. alar. unc. 3—4.

LEPIDOPTERA.

ZEUZERA: cyaneous, with all the wings coloured alike, orange, with cyaneous spots, and a broad central longitudinal bar of the same colour. Expanse of the wings, from 3 to 4 inches.

Donovan observes that "Cramer has given the figure of a small specimen of this fine Phalæna under the trivial name of Mineus. This is evidently of the male; the antennæ of which are not, however, very correctly expressed. Both sexes of this rare insect are represented in the annexed plate, the drawings of which were taken from specimens met with in Bengal, by Mr. Fichtel of Vienna. The originals are at this time in the cabinet of the Emperor of Germany." The same insect, but apparently from a very indifferent specimen, is figured by Eschscholtz in Kotzebue's Entdeckungs Reise in die Sud. See, &c. Weimar, 1821, *p.* 219, *pl.* 11, *fig.* 29, under the name of Zeuzera viridicans.

ZEUZERA SCALARIS.

Plate LIII. fig. 2.

SPECIES. ZEUZERA SCALARIS: nivea; thorace utrinque linea fulva; alis niveis, anticis strigis numerosissimis transversalibus abbreviatis nigris, striaque longitudinali fulva, posticis immaculatis. Expans. alar. unc. $2\frac{1}{4}$.

ZEUZERA: snow-white; the thorax on each side with a fulvous line; wings snow-white, the anterior with many short transverse black lines and a fulvous longitudinal line, the posterior immaculate. Expanse of the wings, $2\frac{1}{4}$ inches.

SYN. Cossus Scalaris, *Fabr. Ent. Syst.* III. 1. *p.* 5. *Mant. Ins.* 2. *p.* 135, (Hepialus s.)

Described by Fabricius, from a specimen in the cabinet of Mr. Monson, as a native of China. The specimen represented in the annexed plate was brought from Bengal, and is, as well as the preceding, at this time in the cabinet of the Emperor of Austria.

LITHOSIA SANGUINOLENTA.

Plate LIII. fig. 3.

FAMILY. LITHOSIIDÆ, *Stephens.*

GENUS. LITHOSIA, *Fabricius.* Callimorpha p. *Latreille.*

SPECIES. LITHOSIA SANGUINOLENTA: alis incumbentibus, niveis, anticis costa sanguinea, per thoracis partem anticam ducta, posticis maculis atris; abdomine fulvo annulis nigris. Expans. alar. $2\frac{1}{3}$ unc.

LEPIDOPTERA.

LITHOSIA: with the wings incumbent, snow-white; the anterior with the front edge sanguineous, which is carried across the front of the thorax; posterior pair spotted with black; abdomen fulvous with black rings. Expanse of the wings, $2\frac{1}{3}$ inches.

SYN. Bombyx sanguinolenta, *Fabr. Ent. Syst.* III. 1. *p.* 473.

Donovan observes that this is a rare species, and that his specimen was received from Bombay.

PSILURA FIGURA.

Plate LIV. fig. 1.

FAMILY. ARCTIIDÆ, *Stephens*. Phalæna, (Bombyx p.) *Linnæus*.
GENUS. PSILURA, *Stephens*. Liparis p. *Ochs*. Laria p. *Schrank*.
SPECIES. PSILURA FIGURA: alis anticis albidis, fusco nigroque maculatis, medio figura 7 nigra notatis; posticis cinereis; abdomine carneo. Expans. alar. unc. $3\frac{1}{8}$.

PSILURA: with the anterior wings whitish, spotted with brown and black, and marked in the middle with a black character like the figure 7; posterior wings cinereous; abdomen pink. Expanse of the wings, $3\frac{1}{8}$ inches.

Donovan states that he received this nondescript species from Madras. As it agrees with our common moth, the black arches (Ps. Monacha) in its general character, and in the peculiar pink colouring of the abdomen, I have placed it in the genus of which that species is the type. In the markings of the wings, however, it seems to make a nearer approach to the tussock moths.

OPHIUSA? STRIGATA.

Plate LIV. fig. 2.

FAMILY. NOCTUIDÆ, *Stephens*. Phalæna, (Noctua) *Linnæus*.
GENUS. OPHIUSA? *Ochs. Tr. Steph.* Ophideres? *Boisduval Faune de l'Océanie, p.* 245.
SPECIES. OPHIUSA STRIGATA: "alis anticis fuscis, litura longitudinali viridi; posticis luteis, lunula limboque nigris." (*Donovan*). Expans. alar. unc. 3.

LEPIDOPTERA.

OPHIUSA: with the "anterior wings brown, with a longitudinal green daub; posterior pair yellowish, with a lunar spot and border of black." Expanse of the wings, 3 inches.

Donovan observes that "this moth agrees entirely with the Noctua Dioscoreæ of Fabricius, (*Ent. Syst. t.* 3, *p.* 2, *p.* 16, *n.* 26.) except in having a large green streak on the upper wings. It appears, indeed, to have been hitherto confounded with that species, either as a sexual difference, or variety; but we are persuaded it is neither. Both sexes of N. Dioscoreæ, in particular, have occurred to our observation, without this streak. Found in Bengal." It should be observed, however, that Fabricius, not only in this, but also in the allied species Materna, (figured by Drury, *vol.* 2, *t.* 13.) states that the colour of the anterior wings is liable to great variation. The emargination of the posterior margin of the fore wings in this figure, is also noticed by Fabricius under Dioscoreæ, and is probably a sexual character.

EREBUS HIEROGLYPHICUS.

Plate LIV. fig. 3.

GENUS. EREBUS, *Latreille.* Thysania, *Dalm.* Noctua, *Fabricius.*

SPECIES. EREBUS HIEROGLYPHICUS: alis dentatis, atris; anticis fascia subapicali abbreviata albida maculaque subocellari; posticis margine bisinuato. Expans. alar. unc. $3\frac{1}{4}$—$\frac{1}{2}$.

EREBUS: with the wings dentated, black; the anterior with a short whitish line near the tip, and a large central ocellated spot; the posterior immaculate with two deep notches. Expanse of the wings, $3\frac{1}{4}$—$\frac{1}{2}$ inches.

SYN. Phalæna (Noctua) hieroglyphica, *Drury App. vol.* 2, *pl.* 2, *fig.* 1. *Oliv. Enc. Méth.* 8. 253. *Fabr. Ent. Syst.* III. 2. p. 11.
Phalæna Magdouia, *Cramer Pap. pl.* 174, *fig.* F.

"Common in the East Indies."—*Donovan.*

Order. NEUROPTERA. *Linnæus.*

MYRMELEON PARDALIS.

Plate LV. fig. 1.

SECTION. PLANIPENNES, *Latreille.* (Filicornes.)
FAMILY. MYRMELEONIDÆ, *Leach.*
GENUS. MYRMELEON, *Linn.* &c.
SPECIES. MYRMELEON PARDALIS : alis albis, maculis plurimis nigris sparsis, pedibus nigris femoribus flavis, corpore flavo nigroque vario. Expans. alar. unc. 3¼.
MYRMELEON : with the wings white, with many black dots scattered over them, especially in the anterior pair ; legs black, with the thighs yellow ; body varied with yellow and black. Expanse of the wings, 3¼ inches.
SYN. Myrmeleon Pardalis, *Fabr. Ent. Syst.* 2. *p.* 92.

MYRMELEON PUNCTATUM.

Plate LV. fig. 2.

SPECIES. MYRMELEON PUNCTATUM : alis hyalino-albidis, venis e punctis niveis nigrisque alternis reticulatis, punctoque magno niveo stigmaticali ; pedibus flavescentibus. Expans. alar. unc. 2¼.
MYRMELEON : with the wings hyaline whitish, with the veins reticulated alternately with black and white, and with a large white stigmal spot, legs yellowish. Expase of the wings, 2¼ inches.
SYN. Myrmeleon Punctatum, *Fabr. Ent. Syst.* 2. *p.* 94.

Donovan observes upon the two preceding species —" Few species of the Myrmeleon genus have been discovered. Linnæus describes only five, and those are chiefly natives of Europe. Fabricius adds seven more, besides three others in the genera Ascalaphus, in the Entomologia Systematica, and particularly two from India, in the cabinet of Sir Joseph Banks, Bart. M. Pardalis and M. Punctatum ; these are the only Indian Myrmeleons hitherto ascertained ; to the Entomologist they are equally interesting as new and unfigured species, but M. Pardalis is much superior in beauty to the other. The characteristic distinction of M. Punctatum, is the alternate black and white specks, or interrupted dashes in the reticulations of the wings. M. Pardalis is

NEUROPTERA.

reticulated also with delicate brownish nerves, its general colour a fine yellow, elegantly barred with transverse streaks of brown; it is from the coast of Coromandel."

The natural history of the Myrmeleon-larva is curious, and has been traced in some of the European kinds by Reaumur, Roesel, and others, particularly in the M. Formica-Leo of Geoffroy; this creature, among other peculiarities, is furnished in the front with a large pair of forceps, with which it takes its prey. It forms circular cavities in the sand, and concealing itself in the centre with only the forceps above the surface, catches the weaker or unwary insects that come within the verge of its cell.

Order. **HYMENOPTERA**, *Linnæus.*

CHRYSIS IMPERIALIS.

Plate LVI. fig. 1.

SECTION. TUBULIFERA, *Latreille.*
FAMILY. CHRYSIDIDÆ, *Leach.*
GENUS. CHRYSIS, *Linnæus.*
SPECIES. CHRYSIS IMPERIALIS: "thorace viridi, fascia cyanea; abdomine antice cyaneo violaceoque fasciato, medio aureo, postice rubro, quadri-dentato."—*Donovan.* Long. Corp. $\frac{3}{8}$ unc.
CHRYSIS: with the thorax green, with a blue band; anterior part of the abdomen blue fasciated with violet, golden in the middle, posterior end red, with four teeth. Length of the body, $\frac{3}{8}$ of an inch.
SYN. Chrysis fasciata, *Donovan*, 1st edit. nec Chrysis fasciata, *Fabr. Syst. Piez. p.* 175.

" This charming insect is from Tranquebar, where we have every reason to believe it to be uncommonly rare. The species does not appear to be described by any author, the only specimen we are acquainted with is in the cabinet of the Right Hon. Sir J. Banks, Bart."—*Donovan.*

This species must not be confounded with the Chrysis fasciata of Fabricius, a

HYMENOPTERA.

South American species, for which it will be convenient to retain the specific name of C. fasciata, (especially as it is now impossible to decide the priority of the employment of the name fasciata between Fabricius or Donovan, both having been published in the year 1804.) I have, therefore, proposed for Donovan's insect a name, which will not be thought inapplicable.

STILBUM OCULATUM.

Plate LVI. fig. 2.

GENUS. STILBUM, *Spinola.* Chrysis p. *Fabricius, Donovan.*
SPECIES. STILBUM OCULATUM : viride, nitens, punctatum; abdomine rotundato, viride, versus apicem utrinque macula ocellata aurea ; ano cæruleo sex-dentato. Long. Corp. unc. ½.
STILBUM : green, shining, punctured ; abdomen rounded above, green, with an eye-like golden spot on each side near the tip ; anus blue with six teeth. Length of the body, ½ an inch.
SYN. Chrysis oculata, *Fabr. Ent. Syst.* 2, *p.* 239. *Sp. Ins. p.* 455. *Syst. Piez. p.* 171.

"Chrysis oculata is distinguished for the peculiar brilliancy of its colours. It was found by Dr. Kœnig at Tranquebar. Fabricius describes this insect from a specimen in the collection of Sir Joseph Banks, Bart.; from whence our figure is also taken. We have since received the same kind from Bengal, through the medium of Mr. Fichtel of Vienna."—*Donovan.*

STILBUM SPLENDIDULUM.

Plate LVI. fig. 3.

SPECIES. STILBUM SPLENDIDULUM : viride, cyaneo nitidum ; abdomine rotundato; ano cæruleo quadri-dentato ; pedibus viridibus, tibiis fuscis. Long. Corp. $\frac{9}{10}$ unc.
STILBUM : green, shining with blue ; abdomen rounded above ; anus blue, with four teeth ; legs green, with brown tibiæ. Length of the body, $\frac{9}{10}$ of an inch.
SYN. Chrysis splendidula, *Fabr. Sp. Ins. p.* 454. *Ent. Syst.* II. *p.* 238. *Syst. Piez. p.* 170.

Donovan states that this species is "very scarce, and that it is a native of Tranquebar, where it was discovered by Dr. Kœnig. Fabricius describes this insect from a specimen in the cabinet of Sir Joseph Banks, Bart. A variety of the same

HYMENOPTERA.

species is found in New Holland." It is most probable that the Australasian specimens constitute a distinct but very closely allied species. The Stilbum princeps of G. R. Gray, figured in Griffith's Animal Kingdom, Insects, *pl.* 77, from the collection of the Rev. F. W. Hope, is from Melville Island, and is perhaps identical with the New Holland variety, mentioned by Fabricius and Donovan.

VESPA CINCTA.

Plate LVII. fig. 1.

Section. Aculeata, *Latreille.*
Family. Vespidæ, *Leach.*
Genus. Vespa, *Linnæus.*
Species. Vespa Cincta : nigra ; capite nigro ; thorace maculato scutelloque obscuro, fulvis ; abdomine atro fascia ferruginea ; alis ferrugineis, basi nigris. Long. Corp. unc. 1.
Vespa : black, with the head black ; thorax with two spots on each side before the wings, and the scutellum obscure fulvous ; abdomen black, with a ferruginous or fulvous bar ; wings ferruginous ; black at the base. Length of the body, 1 inch.
Syn. Vespa cincta, *Fabr. Ent. Syst.* 2. *p.* 253. *Syst. Piez. p.* 253. *St. Fargeau Hist. Nat. Hymenopt.* 1. *p.* 505.
Sphex tropica, *Sulzer. Hist. Ins. tab.* 27, *fig.* 5.

Either this species is subject to considerable variation in its colours, or (which appears to me to be the case,) several distinct species have been confounded together under the name of V. cincta. Fabricius describes V. cincta from Tranquebar with the characters which I have abstracted above, adding a variety from the Cape of Good Hope, and a species under the name of V. affinis, being, as he says, " Nimis affinis V. cinctæ." M. le Comte de Saint Fargeau has also described two other varieties, under the name of cincta, neither of which precisely agree with the Fabrician character.

HYMENOPTERA.

EUMENES PETIOLATA.

Plate LVII. fig. 2.

GENUS. EUMENES, *Latreille.*

SPECIES. EUMENES PETIOLATA : capite ferrugineo fascia verticali nigra ; thorace ferrugineo, antice flavo ; petiolo ferrugineo, basi fasciaque subapicali nigris, utrinque dente parvo armato ; segmento secundo abdominis fascia atra, reliquis flavescentibus. Long. Corp. unc. $1\frac{1}{3}$.

EUMENES : with the head ferruginous, with a black vertical line ; thorax ferruginous, pale yellow in front ; petiole ferruginous, with the base and a band near the tip black armed on each side with a small tooth ; second abdominal segment with a black band, the remaining segments yellowish. Length of the body, $1\frac{1}{8}$ inch.

SYN. Vespa petiolata, *Fabr. Ent. Syst.* 2. p. 278. *Syst. Piez.* p. 284.

Donovan states that "this, with the foregoing species, is remarkably common in many parts of the East Indies. Fabricius speaks of it as a native of Malabar." The Chinese Vespa conica of Fabricius does not appear to be specifically distinct from the present species.

EUMENES ARCUATA.

Plate LVII. fig. 3.

EUMENES ARCUATA : nigra, fronte orbituque oculorum flavis ; thorace flavo maculato et sub scutello flavo, cruce nigra notato ; petiolo elongato incurvo atro, maculis duabus mediis et duabus ante apicem flavis ; segmento secundo abdominis fasciis duabus, reliquis unica, interruptis flavis. Long. Corp. $1\frac{1}{4}$ unc.

EUMENES : black, with the face and the edges of the eyes yellow ; thorax with yellow spots, behind the scutellum yellow with a black cross ; petiole long, incurved, black, with two spots at the middle and two near the tips of yellow ; second abdominal segment with two bands, the remainder with one, all being interrupted in the middle and yellow. Length of the body, $1\frac{1}{8}$ inch.

SYN. Vespa armata, *Fabr. Ent. Syst.* 2. p. 276. *Syst. Piez.* p. 287.

"Described by Fabricius as a native of New Holland; we have received it Madras."—*Donovan.*

HYMENOPTERA.

POLISTES MACAENSIS.

Plate LVII. fig. 4.

GENUS. POLISTES. *Latreille.*

SPECIES. POLISTES MACAENSIS: capite flavo, inter oculos nigro; thorace flavo, dorso obscururo, lineis tribus nigris; scutello flavo bilineato; abdomine flavo strigis tribus undatis. Long Corp. unc. 1.

POLISTES: with the head yellow, black between the eyes; thorax yellow, darker above, with three black longitudinal lines; scutellum yellow, with two transverse lines; abdomen yellow, with three waved black transverse lines. Length of the body, 1 inch.

SYN. Vespa Macaensis, *Fabr. Ent. Syst.* 2. p. 259. *Syst. Piez.* p. 272.

"The specimen from which our figure is copied, is in the cabinet of the Right Hon. Sir J. Banks, Bart., from Macao."—*Donovan.* It is also very common in China, and appears to be the species so frequently represented in the drawings sent to this country, together with its nest, which is attached to the twigs of trees, and is composed of cells without any outer covering.

POLISTES TEPIDA.

Plate LVII. fig. 5.

SPECIES. POLISTES TEPIDA: capite nigro; labro ferrugineo; thorace nigro; collare punctisque duobus magnis dorsalibus ferrugineis; abdominis segmento primo nigro, secundo et tertio nigris, margine ferrugineo, reliquis totis ferrugineis. Long. Corp. $\frac{9}{10}$ unc.

POLISTES: with the head black; upper lip ferruginous; thorax black; collar and two large dorsal spots ferruginous; first segment of abdomen black, second and third black with the margin ferruginous, the remainder entirely ferruginous. Length of the body, $\frac{9}{10}$ths of an inch.

SYN. Vespa tepida, *Fabr Ent. Syst.* 2. p. 262. *Syst. Piez.* p. 271.

Fabricius gives as the habitat of this insect, "Nova Hollandia, *Mus. Dom. Banks.*"

HYMENOPTERA.

XYLOCOPA NASALIS. ♀.

Plate LVII. fig. 6.

FAMILY. APIDÆ, *Leach.*

GENUS. XYLOCOPA, *Fabricius.* Apis p. *Linnæus.*

SPECIES. XYLOCOPA NASALIS: nigra, nitida; subtus pedibusque hirsutissimis; alis nigris, purpureo (ad basin), viridique (ad apicem); nitentibus, facie sub antennas in fœmina alba. Long. Corp. 1¼ unc.

XYLOCOPA: black, shining; under side of the body and the legs very hairy; wings black, shining with purple at the base, and green towards the tips; the nose white in the female. Length of the body, 1¼ inch.

SYN. Xylocopa violacea, *Donovan, 1st edit.*

Donovan appears to have been in doubt as to the propriety of considering this very common Chinese insect as identical with the south of Europe X. violacea, *Linnæus*, with which it has hitherto been confounded. It is, however, much more elongated than that species, and is distinguished by the white nose of the female, whence I have proposed for it the specific name of X. nasalis.

Order. DIPTERA. *Linnæus.*

DIOPSIS INDICA.

Plate LVIII.

FAMILY. MUSCIDÆ, *Leach.*

GENUS. DIOPSIS, *Linnæus.*

SPECIES. DIOPSIS INDICA: ferruginea, oculis, thorace toto, abdomine postice, alarum macula apicali, spinisque scutellaribus nigris. Long. Corp. lin. 4. Expans. alar. lin. 6.

DIOPSIS: ferruginous, with the eyes, thorax, posterior part of the abdomen, terminal spot on the wings, and scutellar spines black. Length of the body, 4 lines. Expanse of the wings, 6 lines.

DIPTERA.

Syn. Diopsis Indica, *Westw. in Trans. Soc. Linn. vol.* 17, *p.* 299.
Diopsis Ichneumonea, *Donovan, 1st edit.*

The appearance of this curious insect is peculiarly striking. Nothing can be more singular than the disposition of the eyes, which are situated at the extremity of two long immoveable pedicles arising from the head, most exactly in that part which in other insects bears the antennæ. In this particular the Diopsis differs not only from other insects of the kindred genera, but also from all the other kinds we are acquainted with. Some few of the Cancri, &c. have indeed the eyes placed at the extremity of elongated pedicles, as is, for example, instanced in the Cancer angulatus, yet these are obviously dissimilar in construction, for they are moveable at the base, and may be directed towards any object, at the will of the animal, with the utmost facility; but to accomplish this, the motion of the pedicle in the Diopsis must be necessarily accompanied by that of the head, or even of the whole body. The eyes of the latter are notwithstanding so conveniently stationed at the globular extremity of the pedicles, as to embrace a far more comprehensive range of sight than is usual with the generality of insects.

To the inexperienced entomologist, the Diopsis would rather seem to be furnished with remarkable horns, and to be destitute of eyes, although the latter are so very conspicuous when they are pointed out; it is, on the contrary, the true horns, or antennæ, that are so minute as to be most likely to escape attention, for each of these consists only of a single setaceous hair, or bristle, seated on a very small tubercle just beneath the eye.

It has been previously intimated, in the observations on the genus Paussus, that the first account of the Diopsis was inserted in a small tract published by Linnæus, at Upsal, in 1775. From this we learn, that both the Diopsis and the Paussus were found by Andreas Dahl, among a parcel of insects in the possession of Dr. Fothergill, of London, by whom they were sent to Linnæus. These consisted chiefly of insects collected in North America and Guinea, but the habitat, either of the Paussus, or the Diopsis in particular, it is very certain was by no means exactly known. Fuessly notwithstanding describes the D. ichneumonea, upon this ambiguous authority only, as a native of Cayenne, and after him Gmelin notes the same insect from South America, and Guinea, perhaps with as little reason. Latreille tells us it is from the coast of Angola, on the information of Perrin, a zealous naturalist of Bordeaux. Donovan's specimens of the insects here figured, and which he says are most assuredly the Diopsis ichneumonea of Linnæus, were brought from Bengal, where they

DIPTERA.

were discovered by Mr. Fichtel, who, he adds, has thus established the habitat of this singular creature beyond dispute.

Linnæus, to whom only an individual species of Diopsis was known, as usual with him under such circumstances, does not assign to it any specific character. Donovan states that he was acquainted with another species of this genus, a native of Africa, in the collection of T. Marsham, Esq. which rendered a deviation from his example excusable, although the latter was at the time undescribed.

Since the preceding observations by Donovan was written, great additions have been made to this curious genus. Fabricius, like Donovan, adopted the Linnæan specific name ichneumonea, but confounded under that name two species distinct from each other, and from the original species. Illiger added another species, D. nigra. Dalman described three new African species; Weidemann another, which he named in honour of Dalman; and Gray another, D. Sykesii, in Griffith's Animal Kingdom. More recently, in the seventeenth volume of the Transactions of the Linnæan Society, I have published a monograph upon the genus, in which I have described thirty-one species, giving coloured figures of the greater portion. All of these species, (with the exception of D. brevicornis of Say, which belongs to a distinct subgenus,) are natives of the old world, inhabiting Guinea, Sierra Leone, Senegal, the Cape of Good Hope, Arabia, and the East Indies.

In this monograph I have endeavoured to prove, that Guinea, or the adjacent parts of Western Africa, is the true locality of D. ichneumonia; whilst Donovan's species, which belongs to a distinct section of the genus with the spot on the wings terminal, (and not merely a sub-apical fascia, as in D. ichneumonia,) I have given as a distinct species, under the name of D. Indica.

ALPHABETICAL INDEX.

In this Index the names employed both in the present and former editions are introduced, in order to render the references which have been made by writers to the former edition available. The names, generic, sub-generic, or specific, first employed in the present edition are distinguished by a *.

*Acanthosoma uniguttata, Pl. 14, f. 5.
*Anthia *6-guttata, Pl. 4, f. 1.
*Anthocharis Danae, Pl. 26, f. 2.
*Anthocaris Eucharis, Pl. 27, f. 4.
*Anthocaris Genutia, Pl. 27, f. 5.
*Argynnis Thyelia, Pl. 31, f. 3.
*Aphana? festiva, Pl. 7, f. 2.
Apis violacea, Pl. 57, f. 6.
*Biblis Hiarbas, Pl. 32, f. 3.
*Brachinus bimaculatus, Pl. 4, f. 2.
Buprestis Ænea, Pl. 3, f. 3.
Buprestis Chrysis, Pl. 3, f. 2.
Buprestis *(Sternocera) Chrysis Pl. 3, f. 2.
Buprestis *(Sphenoptera?) confusa, Pl. 3, f. 3.
Buprestis 4-maculata, Pl. 3, f. 4.
Buprestis *(Anthaxia) 4-maculata, Pl. 3, f. 4.
Buprestis sternicornis, Pl. 3, f. 1.
Buprestis *(Sternocera) sternicornis, Pl. 3, f. 1.
*Calandra palmarum, Pl. 6, f. 2.
Carabus bimaculatus, Pl. 4, f. 2.
Carabus 6-maculatus, Pl. 4, f. 1.
*Castnia Evalthe, Pl. 22.
*Cethosia Cyane, Pl. 35, f. 2.
*Cethosia Cydippe, Pl. 34, f. 1.
Cetonia cærulea, Pl. 2, f. 5.
Cetonia Histrio, Pl. 2, f. 4,
Chrysis fasciata, Pl. 56, f. 1.
Chrysis imperialis, Pl. 56, f. 1.
Chrysis oculata, Pl. 56, f. 2.
Chrysis splendidula, Pl. 56, f. 3.
Cicada Indica, Pl. 8, f. 3.

Cicada *speciosa, Pl. 8, f. 3.
Cimex cruciatus, Pl. 14, f. 3.
Cimex mactans, Pl. 14, f. 4.
Cimex nigripes, Pl. 14, f. 1.
Cimex papillosus, Pl. 14, f. 2.
Cimex serratus, Pl. 14, f. 7.
Cimex uniguttatus, Pl. 14, f. 5.
Cimex viridis, Pl. 14, f. 6.
*Colias *(Callidryas) Scylla, Pl. 28, f. 3.
Curculio palmarum, Pl. 6, f. 2.
Curculio regalis, Pl. 6, f. 1.
*Danais affinis, Pl. 25, f. 2.
Diopsis ichneumonea, Pl. 58.
Diopsis *Indica, Pl. 58.
*Dynastes *(Chalcosoma) Atlas, Pl. 1.
*Erebus hieroglyphicus, Pl. 54, f. 3.
*Erycina Æmulius, Pl. 44, f. 2.
*Erycina (Zemeros) Allica, Pl. 37, f. 2.
*Erycina *Athemon, Pl. 46, f. 3.
*Erycina? Bibulus, Pl. 46, f. 1.
*Erycina? Livius, Pl. 46, f. 4.
*Erycina Lucanus, Pl. 43, f. 4.
*Erycina Ovidius, Pl. 46, f. 7.
*Erycina Petronius, Pl. 43, f. 2.
*Erycina? Plutargus, Pl. 48, f. 3.
*Erycina Ptolomæus, Pl. 46, f. 6.
*Erycina Regulus, Pl. 43, f. 3.
*Erycina Tarquinius, Pl. 44, f. 1.
*Erycina Thucydides, Pl. 43, f. 1.
*Eumenes arcuata, Pl. 57, f. 3.
*Eumenes petiolata, Pl. 57, f. 2.

ALPHABETICAL INDEX.

Fulgora festiva, Pl. 7, fig. 2.
Fulgora hyalinata, Pl. 7, f. 3.
Fulgora lineata, Pl. 8, f. 1.
Fulgora pallida, Pl. 8, f. 2.
Fulgora *pyrorhina, Pl. 7, f. 1.
Fulgora pyrorhynchus. Pl. 7, f. 1.
Gryllus monstrosus, Pl. 12, f. 3.
Gryllus punctatus, Pl. 12, f. 2.
Gryllus reticulatus, Pl. 12, f. 1.
*Goniapteryx Mærula, Pl. 27, f. 1.
*Gymnetis cærulea, Pl. 2, f. 5.
*Gymnopleurus Kœnigii, Pl. 2, f. 3.
*Gymnopleurus, miliaris, Pl. 2, f. 2.
*Hesperia Augias, Pl. 48, f. 1.
*Hesperia Busiris, Pl. 52, f. 2.
*Hesperia Catullus, Pl. 50, f. 4.
*Hesperia Celsus, Pl. 52, f. 3.
*Hesperia Chemnis, Pl. 49, f. 1.
*Hesperia Ennius, Pl. 15, f. 1.
*Hesperia Epictetus, Pl. 48, f. 4.
*Hesperia Galenus, Pl. 50, f. 3.
*Hesperia Jovianus, Pl. 50, f. 1.
*Hesperia *(Eantis) Mithridates, Pl. 42, f. 3.
*Hesperia Numitor, Pl. 44, f. 3.
*Hesperia Orcus, Pl. 52, f. 1.
*Hesperia Origines, Pl. 41, f. 2.
*Hesperia Polybius, Pl. 51, f. 2.
*Hesperia Propertius. Pl. 47, f. 2.
*Hesperia Salvianus, Pl. 50, f. 2.
*Hesperia Spio, Pl. 50, f. 5.
*Hesperia *(Eantis) Thrasibulus, Pl. 49, f. 4.
*Hesperia Thrax, Pl. 49, f. 2.
*Hesperia Tibullus, Pl. 47, f. 3.
*Hesperia Zeleucus, Pl. 51, f. 3.
*Hipparchia Arcesilaus, Pl. 30, f. 2.
*Hipparchia Baldus, Pl. 36, f. 2.
*Hipparchia Crantor, Pl. 37, f. 4.
*Idea agelia, Pl. 24.
*Leptocircus Curius, Pl. 47, f. 1.
Lithosia sanguinolenta, Pl. 53, f. 3.
Locusta Amboinensis, Pl. 13, f. 1.
Locusta citrifolia, Pl. 13, f. 2.
*Locusta *(Phymatea) punctata, Pl. 12, f. 2.
*Locusta *(Monachidia) reticulata, Pl. 12, f. 1.

Mantis Gigas, Pl. 9.
Mantis siccifolia, Pl. 11, f. 1. 2.
Mantis viridis, Pl. 10.
*Morpho Menetho, Pl. 30, f. 1.
Myrmeleon Pardalis, Pl. 55, f. 1.
Myrmeleon punctatum, Pl. 55, f. 2.
*Nymphalis *Ancœa, Pl. 37, f. 3.
*Nymphalis Auge, Pl. 36, f. 4.
*Nymphalis Cænobita, Pl. 35, f. 3.
*Nymphalis *(Aconthea) cocalia, Pl. 36, f. 1.
*Nymphalis cocles, Pl. 23, f. 2.
*Nymphalis Dirce, Pl. 34, f. 2.
*Nymphalis Eribotes, Pl. 33, f. 3.
*Nymphalis Eurinome, Pl. 34, f. 3.
*Nymphalis Fatima, Pl. 31, f. 2.
*Nymphalis Gnidia, Pl. 32, f. 2.
*Nymphalis Hippona, Pl. 35, f. 1.
*Nymphalis Isodore, Pl. 33, f. 4.
*Nymphalis Isis, Pl 33, f. 1.
*Nymphalis Liberia, Pl. 30, f. 4.
*Nymphalis *Lirissa, Pl. 37, f. 5.
*Nymphalis Octavius, Pl. 29, f. 2.
*Nymphalis *Orsis, Pl. 30, f. 3.
*Nymphalis Periander, Pl. 37, f. 1.
*Nymphalis Phegea, Pl. 31, f. 1.
*Nymphalis Philomela, Pl. 25, f. 3.
*Nymphalis Phorcys, Pl. 33, f. 2.
*Nymphalis *(Charaxes) Pyrrhus, Pl. 29, f. 3.
*Nymphalis *(Charaxes) Tiridates, Pl. 23, f. 3.
*Onthopagus spinifex, Pl. 2, f. 1.
*Ophuisa strigata, Pl. 54, f. 2.
*Ornithopterus Heliacon, Pl. 19, f. 1.
*Ornithopterus Priamus, Pl. 16.
*Ornithopterus *Remus, Pl. 18.
Pausus denticornis, Pl. 5, f. 1.
Pausus Fichtelii, Pl. 5, f. 3.
Pausus pilicornis, Pl. 5, f. 4.
Pausus thoracicus, Pl. 5, f. 2.
*Paussus Fichtelii, Pl. 5, f. 3.
*Paussus pilicornis, Pl. 5, f. 4.
*Paussus thoracicus, Pl. 5, f. 2.
Papilio Achæus, Pl. 41, f. 4.
Papilio Æmulius, Pl. 44, f. 2.
Papilio Æolus, Pl. 42, f. 1.

ALPHABETICAL INDEX.

Papilio Affinis, Pl. 25, f. 2.
Papilio Allica, Pl. 37, f. 2.
Papilio Amaryllis, Pl. 28, f. 1.
Papilio Antenor, Pl. 15, f. 1.
Papilio Antiphus, Pl. 15, f. 2.
Papilio Arcesilaus, Pl. 30, f. 2.
Papilio Astyanax, Pl. 20, f. 1.
Papilio Auge, Pl. 36, f. 4.
Papilio Augias, Pl. 48, f. 1.
Papilio Baldus, Pl. 36, f. 2.
Papilio Bibulus, Pl. 46, f. 1.
Papilio Blandina, Pl. 30, f. 3.
Papilio Busiris, Pl. 52, f. 2.
Papilio Cacta, Pl. 29, f. 1.
Papilio Cænobita, Pl. 35, f. 3.
Papilio Cœnus, Pl. 46, f. 3.
Papilio Castalia, Pl. 28, f. 2.
Papilio Catullus, Pl. 50, f. 4.
Papilio Celsus, Pl. 52, f. 3.
Papilio Chemnis, Pl. 49, f. 1.
Papilio Chiton, Pl. 39, f. 1.
Papilio Cocalia, Pl. 36, f. 1.
Papilio Cocles, Pl. 23, f. 2.
Papilio Crantor, Pl. 37, f. 4.
Papilio Curius, Pl. 47, f. 1.
Papilio Cyane, Pl. 35, f. 2.
Papilio Cydippe, Pl. 34, f. 1.
Papilio Danae, Pl. 26, f. 2.
Papilio Deiphobus, Pl. 17, f. 2.
Papilio Dirce, Pl. 34, f. 2.
Papilio Empedocles, Pl. 17, f. 1.
Papilio Ennius, Pl. 51, f. 1.
Papilio Epictetus, Pl. 48, f. 4.
Papilio Eribotes, Pl. 33, f. 3.
Papilio Eucharis, Pl. 27, f. 4.
Papilio Eurinome, Pl. 34, f. 3.
Papilio Evalthe, Pl. 22.
Papilio Fatima, Pl. 31, f. 2.
Papilio Florus, Pl. 39, f. 4.
Papilio Galenus, Pl. 50, f. 3.
Papilio Genutia, Pl. 27, f. 5.
Papilio Gnidia, Pl. 32, f. 2.
Papilio Heliacon, Pl. 19, f. 1.
Papilio Herodotus, Pl. 39, f. 2.

Papilio Hiarba, Pl. 32, f. 3.
Papilio Hippia, Pl. 25, f. 1.
Papilio Hippocrates, Pl. 45, f. 3.
Papilio Hippona, Pl. 35, f. 1.
Papilio Hylax, Pl. 46, f. 2.
Papilio Idæus, Pl. 19, f. 2.
Papilio Idea, Pl. 24.
Papilio Isidore, Pl. 33, f. 4.
Papilio Isis, Pl. 33, f. 1.
Papilio Jarbas, Pl. 40, f. 3.
Papilio Jovianus, Pl. 50, f. 1.
Papilio Judith, Pl. 27, f. 2.
Papilio Lacedemon, Pl. 17, f. 3.
Papilio Lethe, Pl. 23, f. 1.
Papilio Leucippe, Pl. 26, f. 1.
Papilio Liberia, Pl. 30, f. 4.
Papilio Libythea, Pl. 27, f. 3.
Papilio Liria, Pl. 37, f. 5.
Papilio Lisias, Pl. 40, f. 1.
Papilio Livius, Pl. 46, f. 4.
Papilio Lucanus, Pl. 43, f. 4.
Papilio Mærula, Pl. 27, f. 1.
Papilio Melibœus, Pl. 41, f. 1.
Papilio Menetho, Pl. 30, f. 1.
Papilio Mithridates, Pl. 49, f. 3.
Papilio Nero, Pl. 32, f. 1.
Papilio Numitor, Pl. 44, f. 3.
Papilio Obrinus, Pl. 37, f. 3.
Papilio Octavius, Pl. 29, f. 2.
Papilio Orcus, Pl. 52, f. 1.
Papilio Origines, Pl. 48, f. 2.
Papilio Ovidius, Pl. 46, f. 7.
Papilio Pann, Pl. 38, f. 1.
Papilio Panthous, Pl. 18.
Papilio Parrhasius, Pl. 45, f. 5.
Papilio Periander, Pl. 37, f. 1.
Papilio Pericles, Pl. 42, f. 4.
Papilio Petronius, Pl. 43, f. 2.
Papilio Phegea, Pl. 31, f. 1.
Papilio Philippus, Pl. 42, f 3.
Papilio Philomela, Pl. 25, f. 3.
Papilio Phorbas, Pl. 41, f. 5.
Papilio Phorcys, Pl. 33, f. 2.
Papilio Pindarus, Pl. 38, f. 2.

O

ALPHABETICAL INDEX.

Papilio Plato, Pl. 45, f. 2.
Papilio Plinius, Pl. 45, f. 1.
Papilio Plutargus, Pl. 48, f. 3.
Papilio Polybius, Pl. 51, f. 2.
Papilio Polymnestor, Pl. 20, f. 2.
Papilio Priamus, Pl. 16.
Papilio Propertius, Pl. 47, f. 2.
Papilio Ptolomæus, Pl. 46, f. 6.
Papilio Pyrrhus, Pl. 29, f. 3.
Papilio Pythagoras, Pl. 39, f. 3.
Papilio Regulus, Pl. 43, f. 3.
Papilio Romulus, Pl. 46, f. 5.
Papilio Salvianus, Pl. 50, f. 2.
Papilio Scylla, Pl. 28, f. 3.
Papilio Sophia, Pl. 36, f. 3.
Papilio Sophocles, Pl. 40, f. 2.
Papilio Spio, Pl. 50, f. 5.
Papilio Strephon, Pl. 42, f. 2.
Papilio Tarquinius, Pl. 44, f. 1.
Papilio Thales, Pl. 40, f. 4.
Papilio Theocritus, Pl. 45, f. 4.
Papilio Thucydides, Pl. 43, f. 1.
Papilio Thyelia, Pl. 31, f. 3.
Papilio Tibullus, Pl. 47, f. 3.
Papilio Tiridates, Pl. 23, f. 3.
Papilio Thrasibulus, Pl. 49, f. 4.
Papilio Thrax, Pl. 49, f. 2.
Papilio Tyrtæus, Pl. 41, f. 2.
Papilio Ulysses, Pl. 21.
Papilio Vulcanus, Pl. 38, f. 3.
Papilio Xenophon, Pl. 41, f. 3.
Papilio Zeleucus, Pl. 51, f. 3.
*Pentatoma cruciata, Pl. 14, f. 3.
*Pentatoma mactans, Pl. 14, f. 4.
*Phasma *(Platycrana) Edule, Pl. 10.
*Phasma *(Cyphocrana) gigas, Pl. 9.
*Phyllium siccifolium, Pl. 11, f. 1. 2.
*Phyllophora Amboinensis, Pl. 13, f. 1.
*Phyllophora citrifolia, Pl. 13, f. 2.
Phalæna figura, Pl. 54, f. 1.
Phalæna hieroglyphica, Pl. 54, f. 3.
Phalæna Mineas, Pl. 53, f. 1.
Phalæna sanguinolenta, Pl. 53, f. 3.
Phalæna scalaris, Pl. 53, f. 2.

Phalæna strigata, Pl. 54, f. 2.
*Pieris Amaryllis, Pl. 28, f. 1.
*Pieris Castalia, Pl. 28, f. 2.
*Pieris Hippia, Pl. 25, f. 1.
*Pieris Judith, Pl. 27, f. 2.
*Pieris *(Iphias) Leucippe, Pl. 26, f. 1.
*Pieris Libythea, Pl. 27, f. 3.
*Pieris Nero, Pl. 32, f. 1.
*Platyrhophalus denticornis, Pl. 5, f. 1.
*Polistes Macaensis, Pl. 57, f. 4.
*Polistes tepida, Pl. 57, f. 5.
*Polyommatus Florus, Pl. 39, f. 4.
*Polyommatus Hippocrates, Pl. 45, f. 3.
*Polyommatus Hylax, Pl. 46, f. 2.
*Polyommatus Parrhasius, Pl. 45, f. 5.
*Polyommatus Plato, Pl. 45, f. 2.
*Polyommatus Plinius, Pl. 45, f. 1.
*Prepodes regalis, Pl. 6, f. 1.
*Pseudaphana hyalinata, Pl. 7, f. 3.
*Pseudaphana pallida, Pl. 8, f. 2.
Psilura figura, Pl. 54, f. 1.
*Raphigaster *incarnatus, Pl. 14, f. 1.
*Rhynchocoris hamata, Pl. 14, f. 7.
*Rhynchocoris viridis, Pl. 14, f. 6.
Scarabæus Atlas, Pl. 1.
Scarabæus Kœnigii, Pl. 2, f. 3.
Scarabæus miliaris, Pl. 2, f. 2.
Scarabæus spinifex, Pl. 2, f. 1.
*Schizodactyla monstrosa, Pl. 12, f. 3.
*Stilbum oculatum, Pl. 56, f. 2.
*Stilbum splendidulum, Pl. 56, f. 3.
*Tesseratoma papillosa, Pl. 14, f. 2.
*Thecla Achæus, Pl. 41, f. 4.
*Thecla Æolus, Pl. 42, f. 1.
*Thecla Chiton, Pl. 39, f. 1.
*Thecla Herodotus, Pl. 39, f. 2.
*Thecla *Isocrates, Pl. 38, f. 1.
*Thecla Jarbas, Pl. 40, f. 3.
*Thecla Lisias, Pl. 40, f. 1.
*Thecla Melibœus, Pl. 41, f. 1.
*Thecla Pericles, Pl. 42, f. 4.
*Thecla Philippus, Pl. 42, f. 3.
*Thecla Phorbas, Pl. 41, f. 5.
*Thecla Pindarus, Pl. 38, f. 2.

ALPHABETICAL INDEX.

*Thecla Pythagoras, Pl. 39, f. 3.
*Thecla Romulus, Pl. 46, f. 5.
*Thecla Strephon, Pl. 42, f. 2.
*Thecla Sophocles, Pl. 40, f. 2.
*Thecla Thales, Pl. 40, f. 4.
*Thecla Theocritus, Pl. 45, f. 4.
*Thecla Tyrtæus, Pl. 41, f. 2.
*Thecla Vulcanus, Pl. 38, f. 3.
*Thecla Xenophon, Pl. 41, f. 3.
*Vanessa cacta, Pl. 29, f. 1.

*Vanesa Lethe, Pl. 23, f. 1.
*Vanessa Sophia, Pl. 36, f. 3.
Vespa arcuata, Pl. 57, f. 3.
Vespa cincta, Pl. 57, f. 1.
Vespa Macaensis, Pl. 57, f. 4.
Vespa petiolata, Pl. 57, f. 2.
Vespa tepida, Pl. 57, f. 5.
*Xylocopa *nasalis, Pl. 57, f. 6.
Zeuzera Minea, Pl. 53, f. 1.
Zeuzera Scalaris, Pl. 53, f. 2.

SYSTEMATIC INDEX.

INSECTA.

I.—MOUTH WITH JAWS.

Order. COLEOPTERA

Family. CARABIDÆ.

Anthia 6-guttata, Pl. 4, f. 1.
Brachinus bimaculatus, Pl. 4, f. 2.

Family. SCARABÆIDÆ.

Onthophagus spinifex, Pl. 2, f. 1.
Gymnopleurus miliaris, Pl. 2, f. 2.
Gymnopleurus Kœnigii, Pl. 2, f. 3.

Family. DYNASTIDÆ.

Dynastes (Chalcosoma) Atlas, Pl. 1.

Family. CETONIIDÆ.

Cetonia Histrio, Pl. 2, f. 4.
Gymnetis cærulea, Pl. 2, f. 5.

Family. BUPRESTIDÆ.

Buprestis (Sternocera) sternicornis, Pl. 3, f. 1.
Buprestis (Sternocera) chrysis, Pl. 3, f. 2.
Buprestis (Sphenoptera?) confusa, Pl. 3, f. 3.
Buprestis (Anthaxia) 4-maculata, Pl. 3, f. 4.

Family. PAUSSIDÆ.

Platyrhopalus denticornis, Pl. 5, f. 1.
Paussus thoracicus, Pl. 5, f. 2.
Paussus Fichtelii, Pl. 5, f. 3.
Paussus pilicornis, Pl. 5, f. 4.

Family. CURCULIONIDÆ.

Prepodes regalis, Pl. 6, f. 1.
Calandra palmarum, Pl. 6, f. 2.

Order. ORTHOPTERA.

Family. PHASMIDÆ.

Phasma (Cyphocrana) Gigas, Pl. 9.
Phasma (Platycrana) Edule ♀, Pl. 10.
Phyllium siccifolium, Pl. 11.

Family. LOCUSTIDÆ.

Locusta (Monachidia) reticulata, Pl. 12, f. 1.
Locusta (Phymatea) punctata, Pl. 12, f. 2.

Family. ACHETIDÆ.

Schizodactyla monstrosa, Pl. 12, f. 3.

Family. GRYLLIDÆ.

Phyllophora Amboinensis, Pl. 13, f. 1.
Phyllophora citrifolia, Pl. 13, f. 2.

Order. NEUROPTERA.

Family. MYRMELEONIDÆ.

Myrmeleon pardalis, Pl. 55, f. 1.
Myrmeleon punctatum, Pl. 55, f. 2.

Order. HYMENOPTERA.

Family. CHRYSIDIDÆ.

Chrysis imperialis, Pl. 56, f. 1.
Stilbum oculatum, Pl. 56, f. 2.
Stilbum splendidulum, Pl. 56, f. 3.

Family. VESPIDÆ.

Vespa cincta, Pl. 57, f. 1.
Eumenes petiolata, Pl. 57, f. 2.
Eumenes arcuata, Pl. 57, f. 3.
Polistes Macaensis, Pl. 57, f. 4.
Polistes tepida, Pl. 57, f. 5.
Xylocopa nasalis, Pl. 57, f. 6.

SYSTEMATIC INDEX.

II.—MOUTH SUCTORIAL.

Order. LEPIDOPTERA.

Family PAPILIONIDÆ.

Ornithopterus Priamus, Pl. 16.
Ornithopterus Remus, Pl. 18.
Ornithopterus Heliacon, Pl. 19, f. 1.
Papilio Antenor, Pl. 15, f. 1.
Papilio Antiphus, Pl. 15, f. 2.
Papilio Empedocles, Pl. 17, f. 1.
Papilio Deiphobus, Pl. 17, f. 2.
Papilio Lacedemon, Pl. 17, f. 3.
Papilio Idæus, Pl. 19, f. 2.
Papilio Astyanax, Pl. 20, f. 1.
Papilio Polymnestor, Pl. 20, f. 2.
Papilio Ulysses, Pl. 21.
Leptocircus Curius, Pl. 47, f. 1
Pieris Hippia, Pl. 25, f. 1.
Pieris Judith, Pl. 27, f. 2.
Pieris Libythea, Pl. 27, f. 3.
Pieris Amaryllis, Pl. 28, f. 1.
Pieris Castalia, Pl. 28, f. 2.
Pieris Nero, Pl. 32, f. 1.
Pieris (Iphias) Leucippe, Pl. 26, f. 1.
Anthocaris Danae, Pl. 26, f. 2.
Anthocaris Eucharis, Pl. 27, f. 4.
Anthocaris Genutia, Pl. 27, f. 5.
Goniapteryx Mærula, Pl. 27, f. 1.
Colias (Callidryas) Scylla, Pl. 28, f. 3.

Family. HELICONIIDÆ.

Idea Agelia, Pl. 24.
Danais affinis, Pl. 25, f. 2.

Family. NYMPHALIDÆ.

Nymphalis Philomela, Pl. 25, f. 3.
Nymphalis Octavius, Pl. 29, f. 2.
Nymphalis Orsis, Pl. 30, f. 3.
Nymphalis Liberia, Pl. 30, f. 4.
Nymphalis ? Phegea, Pl. 31, f. 1.
Nymphalis Fatima, Pl. 31, f. 2.
Nymphalis Gnidia, Pl. 32, f. 2.
Nymphalis Isis, Pl. 33, f. 1.
Nymphalis Phorcys, Pl. 33, f. 2.
Nymphalis Eribotes, Pl. 33, f. 3.
Nymphalis Isidore, Pl. 33, f. 1.
Nymphalis Dirce, Pl. 34, f. 2.
Nymphalis Eurinome, Pl. 34, f. 3.
Nymphalis Hippona, Pl. 35, f. 1.
Nymphalis Cænobita, Pl. 35, f. 2.
Nymphalis Auge, Pl. 36, f. 4.
Nymphalis Periander, Pl. 37, f. 1.
Nymphalis Cocles, Pl. 23, f. 2.
Nymphalis Ancæa, Pl. 37, f. 3.
Nymphalis Lirissa, Pl. 37, f. 5.
Nymphalis (Charaxes) Tiridates, Pl. 23, f. 3.
Nymphalis (Charaxes) Athamas, Pl. 29, f. 3.
Nymphalis (Aconthea) Cocalia, Pl. 36, f. 1.
Vanessa Cacta, Pl. 29, f. 1.
Vanessa Lethe, Pl. 23, f. 1.
Vanessa Sophia, Pl. 36, f. 3.
Cethosia Cydippe, Pl. 34, f. 1.
Cethosia Cyane, Pl. 35, f. 2.
Morpho Menetho, Pl. 30, f. 1.
Biblis Hiarbas, Pl. 32, f. 3.
Argynnis Thyelia, Pl. 31, f. 3.
Hipparchia ? Arcesilaus, Pl. 30, f. 2.
Hipparchia Baldus, Pl. 36, f. 2.
Hipparchia Crantor, Pl. 37, f. 4.

Family. LYCÆNIDÆ.

Erycina (Zemeros) Allica, Pl. 37, f. 2.
Erycina Thucydides, Pl. 43, f. 1.
Erycina Petronius, Pl. 43, f. 2.
Erycina Regulus, Pl. 43, f. 3.
Erycina Lucanus, Pl. 43, f. 4.
Erycina Tarquinius, Pl. 44, f. 1.
Erycina Æmulius, Pl. 44, f. 2.
Erycina Athemon, Pl. 46, f. 3.
Erycina Ptolomæus, Pl. 46, f. 6.
Erycina Ovidius, Pl. 46, f. 7.
Erycina ? Bibulus, Pl. 46, f. 1.
Erycina ? Livius, Pl. 46, f. 4.
Erycina ? Plutargus, Pl. 48, f. 3.
Polyommatus Florus, Pl. 39, f. 4.
Polyommatus Plinius, Pl. 45, f. 1.
Polyommatus Plato, Pl. 45, f. 2.

Polyommatus Hippocrates, Pl. 45, f. 3.
Polyommatus Parrhasius, Pl. 45, f. 5.
Polyommatus Hylax, Pl. 46, f. 2.
Thecla Isocrates, Pl. 38, f. 1.
Thecla Piudarus, Pl. 38, f. 2.
Thecla Vulcanus, Pl. 38, f. 3.
Thecla Chiton, Pl. 39, f. 1.
Thecla Herodotus, Pl. 39, f. 2.
Thecla Pythagoras, Pl. 39, f. 3.
Thecla Lisias, Pl. 40, f. 1.
Thecla Sophocles, Pl. 40, f. 2.
Thecla Jarbas, Pl. 40, f. 3.
Thecla Thales, Pl. 40, f. 4.
Thecla Melibœus, Pl. 41, f. 1.
Thecla Tyrtæus, Pl. 41, f. 2.
Thecla Xenophon, Pl. 41, f. 3.
Thecla Achæus, Pl. 41, f. 4.
Thecla Phorbas, Pl. 41, f. 5.
Thecla Æolus, Pl. 42, f. 1.
Thecla Strephon, Pl. 42, f. 2.
Thecla Philippus, Pl. 42, f. 3.
Thecla Pericles, Pl. 42, f. 4.
Thecla Theocritus, Pl. 45, f. 4.
Thecla Romulus, Pl. 46, f. 5.

Family. HESPERIIDÆ.

Hesperia Numitor, Pl. 44, f. 3.
Hesperia Propertius, Pl. 47, f. 2.
Hesperia Tibullus, Pl. 47, f. 3.
Hesperia Augias, Pl. 48, f. 1.
Hesperia Origines, Pl. 48, f. 2.
Hesperia Epictetus, Pl. 48, f. 4.
Hesperia Chemnis, Pl. 49, f. 1.
Hesperia Thrax, Pl. 49, f. 2.
Hesperia Jovianus, Pl. 50, f. 1.
Hesperia Salvianus, Pl. 50, f. 2.
Hesperia Galenus, Pl. 50, f. 3.
Hesperia Catullus, Pl. 50, f. 4.
Hesperia Spio, Pl. 50, f. 5.
Hesperia Ennius, Pl. 51, f. 1.
Hesperia Polybius, Pl. 51, f. 2.
Hesperia Zeleucus, Pl. 51, f. 3.
Hesperia Orcus, Pl. 52, f. 1.
Hesperia Busiris, Pl. 52, f. 2.

Hesperia Celsus, Pl. 52, f. 3.
Hesperia (Eantis) Mithridates, Pl. 49, f. 3.
Hesperia (Eantis) Thrasibulus, Pl. 49, f. 4.

Family. CASTNIIDÆ.

Castnia Evalthe, Pl. 22.

Family. HEPIALIDÆ.

Zeuzera Minea, Pl. 53, f. 1.
Zeuzera Scalaris, Pl. 53, f. 2.

Family. LITHOSIIDÆ.

Lithosia sanguinolenta, Pl. 53, f. 3.

Family. ARCTIIDÆ.

Psilura figura, Pl. 54, f. 1.

Family. NOCTUIDÆ.

Ophideres? strigata, Pl. 54, f. 2.
Erebus hieroglyphicus, Pl. 54, f. 3.

Order. HEMIPTERA.

Family. PENTATOMIDÆ.

Raphigaster incarnatus, Pl. 14, f. 1.
Pentatoma, cruciata, Pl. 14, f. 2.
Pentatoma mactans, Pl. 14, f. 4.
Acanthosoma uniguttata, Pl. 14, f. 5.
Rhynchocoris viridis, Pl. 14, f. 6.
Rhynchocoris hamata, Pl. 14, f. 7.
Tesseratoma papillosa, Pl. 14, f. 2.

Sub-order. HOMOPTERA.

Family. FULGORIDÆ.

Fulgora pyrorhina, Pl. 7, f. 1.
Fulgora lineata, Pl. 8, f. 1.
Alphana? festiva, Pl. 7, f. 2.
Pseudaphana hyalinata, Pl. 7, f. 3.
Pseudaphana pallida, Pl. 8, f. 2.

Family. CICADIDÆ.

Cicada speciosa, Pl. 8, f. 3.

Order. DIPTERA.

Family. MUSCIDÆ.

Diopsis Indica, Pl. 58.

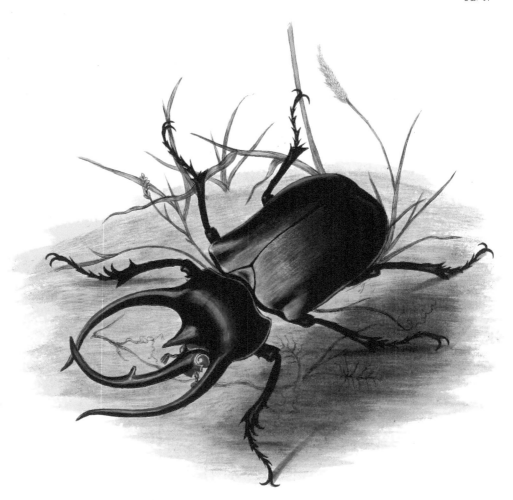

Dynastes (Chalcosoma) Atlas.

PL. I.

1. Onthophagus Spinifex. 2. Gymnopleurus miliaris. 3 &
4. Cetonia Histrio. 5. Gymnetis cærulea.

1. Buprestis sternicornis. 2. Buprestis chrysis.
3. B. confusa. 4. B. 4. maculata.

1. Anthia 6-guttata. 2. Brachinus 2-maculatus

1. Platyrhopalus œnticornis. 2. Paussus thoracicus.
3. P. Fichtelii 4. P. filicornis

1. Prepodes Regalis.   2. Calandra Palmarum.

1. Fulgora pyrorhina 2. Aphana festiva
3. Pseudaphana hyalinata

PL. VIII

1. Fulgora lineata. 2. Pseudaphana pallida.
3. Cicada speciosa.

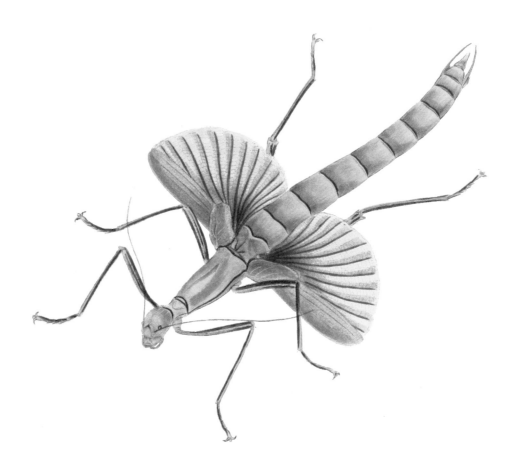

Phasma Edule

PL. XI

1. 2. *Phyllium siccifolium* 3. *Ph. Demerariensis*

PL. XII.

Schizodactyla monstrosa. 1. Locusta reticulata.
2. Locusta punctata.

PL XIII

1. Phyllophora Amboinensis. 2. Ph. stipula.

PL. XIV

1. Raphigaster incarnatus. 2. Tesseratoma papulosa. 3. Pentatoma cruciata.
4. P. mactans. 5. Acanthosoma uniguttata. 6. Rhynchocoris viridis.
7. Rh. hamata.

PL. XV

1. Papilio Antenor.   2. Papilio Antiphus.

PL. XVI

Ornithoptera Priamus

PL. XVII

1. Papilio Empedocles. 2. Papilio Deiphobus.
3. Papilio Lacedemon.

PL. XVIII

Ornithoptera Remus.

PL. XIX

1. Ornithoptera Heliacon. 2. Papilio Idæus

2. Papilio Astyanax. 1. Papilio Polymnestor.

Papilio Ulysses.

PL. XXII

Castnia Evalthe.

Pl. XXIII

1. Vanessa Lethe. 2. Nymphalis Cocles.
3. N. (Charaxes) Tiridates.

PL XXIV

Idea Agelia

1. Pieris Hippia. 2. Danais Affinis.
3. Nymphalis Philomela.

1. Pieris Leucippe. 2. Anthocharis Danæ.

1. Goniapteryx Marula. 2. Pieris Judith. 3. P. Libythea.
4. ♀ Anthocharis Eucharis. 5. A. Genutia.

1. Pieris Amaryllis.   2. Pieris Castalia.
3. Colias Scylla.

PL. XXIX.

1. Vanessa Cacta. 2. Nymphalis Octavius.
3. Nymphalis Athamas.

1. Morpho Menetho. 2. Hipparchus? Brasilianus.
3. Nymphalis orsis. 4. N. Libera.

Pl. XXXI

1. *Nymphalis? Phegea.* 2. *N. Fatima.* 3. *Argynnis Thyelia.*

1. Pieris Nercyl. 2. Nymphalis Guidie.
3. Biblis Hiarbas.

PL. XXXIII

1. Nymphalis Isis. 2. Nymphalis Phorcys.
3. N. Eribotes 4. N. Isidore.

1. Cethosia Cydippe. 2. Nymphalis Dirce. 3. Nymphalis Euremimene.

1. Nymphalis Hippona. 2. Cethosia Cyane.
3. Nymphalis Coenobita.

1. Nymphalis Cocalia. 3. Vanessa Sophia.
2. Hipparchia Baldus. 4. Nymphalis Auge.

1. Nympholis Periander.   2. Erycina Alica.
3. N. Ancæa.   4. Hipparchia Crantor.   5. N. Liris.

1. Thecla Isocrates. 2. Thecla Pindarus.
3. Thecla Vulcanus.

1. Thecla Chiton. 2. Th. Herodotus. 3. Th. Pythagoras. 4. Polyommatus Florus.

1. Thecla Lisias. 2. Thecla Sophocles.
3. Thecla Jarbas. 4. Thecla Thales.

1. Thecla Melibœus 2. Th. Tyrtæus 3. Th. Xenophon.
4. Th. Achæus 5. Th. Phorbas

1. Thecla Æolus.   2. Thecla Erophos.
3. Thecla Phnippus. 4. Thecla Porudes.

1. Erycina Thucidides. 2. Erycina Petronius.
3. Er. Regulus 4. Er. Lucanus.

PL. XLIV

1. *Erycina Tarquinius.*   2. *Erycina Æmulius.*
3. *Hesperia Numitor.*

1. Polyommatus Plinius. 2. P. Plato. 3. P. Hippocrates.
4. Thecla Theocritus. 5. P. Parrhasius.

1. Erycina? Bibulus 2. Polyommatus Hylax 3. Erycina Athemon
4. E. Livius 5. Thecla Romulus 6. E. Ptolomæus 7. E. Ovidius

PL. XLVII

1. Leptocircus Curius. 2. Hesperia Propertius.
3. Hesperia Tibullus.

PL XLVIII

1. Hesperia Augias. 2. Hesperia Origines.
3. Erycina? Plutargus 4. Hesperia Epictetus.

PL. XLIX.

1. Hesperia Chemnis. 2. Hesperia Thrax.
3. H. Mithridates. 4. H. Thrasibulus.

1. Hesperia jovianus. 2. H. Sabrinus. 3. H. Galenus.
4. H. Catullus. 5. H. Spio.

PL. LI.

1. Hesperia Ennius. 2. Hesperia Polybius.
3. Hesperia Tericus.

1. Hesperia Orcus. 2. Hesperia Busiris.
3. Hesperia Celsus.

1. Zeuzera Mincas. 2. Zeuzera scalaris.
3. Lithosia sanguinolenta.

1. Psilura figura. 2. Ophiusa Strigata.
3. Erebus hieroglyphicus.

Pl. LV

1. Myrmeleon Pardalis.   2. Myrmeleon punctatum.

PL. LVI

1. Chrysis imperialis.
2. Stilbum oculatum. 3. Stilbum splendidum.

1. Vespa cincta. 2. Eumenes petiolata. 3. E. arcuata.
4. Polistes Macaensis. 5. P. tepida. 6. Xylocopa nasalis.

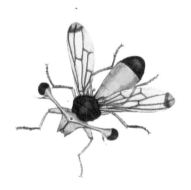

Diopsis Indica